The Horse of the Americas

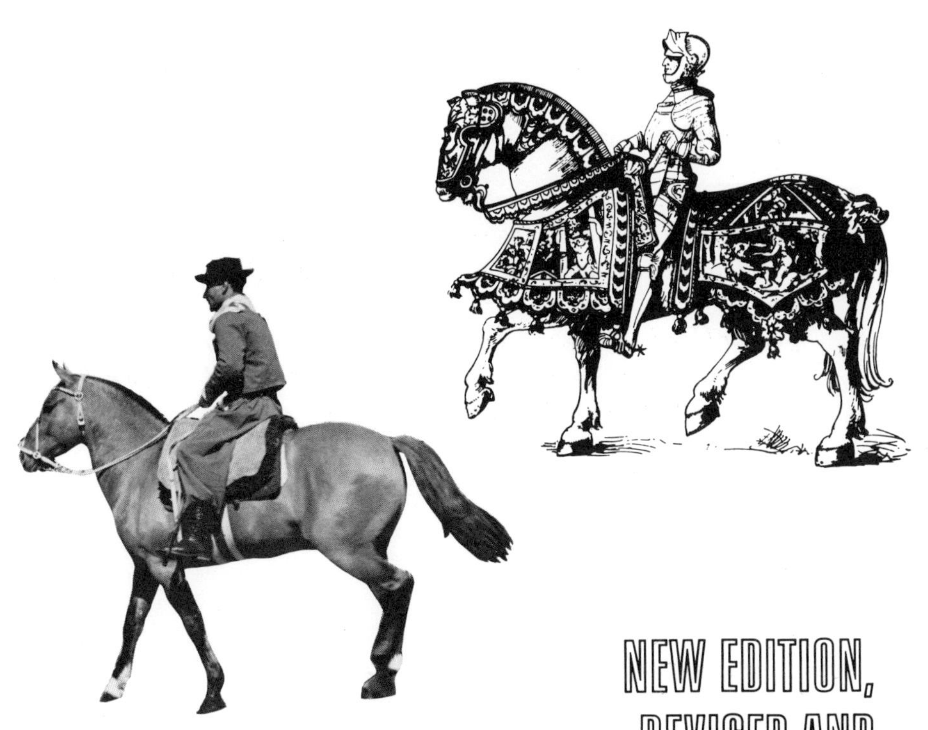

NEW EDITION, REVISED AND ENLARGED

The Horse of the Americas

ROBERT M. DENHARDT

By Robert Moorman Denhardt

The Quarter Horse (3 volumes, Amarillo, 1941–50)
The Horse of the Americas (Norman, 1947, 1975)
Horses of the Conquest (editor), by R. B. Cunninghame Graham (Norman, 1949)
Quarter Horses: A Story of Two Centuries (Norman, 1967)
The King Ranch Quarter Horses: And Something of the Ranch and the Men That Bred Them (Norman, 1970)

Library of Congress Cataloging in Publication Data
Denhardt, Robert Moorman, 1912–
The horse of the Americas.
"Bibliographical sketch": p.
1. Horses—America. 2. Horses—History. 3. Horse breeding—America. 4. Horse breeds. I. Title.
SF284.A6D4 1975 636.1'3'091812 74-5955
ISBN 0-8061-1213-1

 Composed and printed at Norman, Oklahoma, by the University of Oklahoma Press. First edition, 1947; second printing, 1948; third printing, 1949. First printing of the new edition, 1975.

"El caballo criollo, lo sabe Diós, es buen caballo al segundo dia, sin que haya probado alimentos ni bibido nada el primero, y se no lo sabe Diós, lo sabe el diablo, pues es muy viejo."—R. B. Cunninghame Graham, 1890.

Foreword

"In Texas the history of the horse is equally as important as that of its owner." Thus wrote a German in a book of little consequence published in 1886. Discounting intellect in the way that United States voters habitually discount it, the observation might have been made with equal justice of all latitudes in the Americas stocked with Spanish breeds of horses and cattle. Historians call that period of time after the Plains Indians acquired horses and became horsemen their Age of Horse Culture. For English-speakers who ranged the trans-Mississippi West and for Spanish-speakers who ranged on south even below W. H. Hudson's magic land of *Far Away and Long Ago,* there was also an Age of Horse Culture. It lasted until barbed wire and railroads killed both the freedom and necessity of mobility by the horse. While it lasted, men who belonged to the Age said truly, "A man on foot is no man at all," and "A man is no better than his horse." When automobiles and other exponents of the Industrial Age tied all parts of all continents together and made the interdependence between Nevada sage brush and pigeons on Boston Commons palpable to everybody but the most ignorant provincials, the epoch of Horse Culture was definitely past.

The horse still has his uses, and there is no diminution in the pleasure and delight that this noble animal affords human beings. In the West particularly, where the Age of Horse Culture belonged almost exclusively to men, women increasingly share in the pleasure. In the summer time more "range riders" ride over the more than two million acres comprising Yellowstone National Park—a model of civilized administration—than ride over ten million acres of Wyoming ranges to the east of the Park. A majority of these "range riders" are girls and women. The ascendancy of the luxury horse is exemplified in the rage for Palominos, for which few cowmen of rawhide tradition would furnish grass.

The ideas and institutions behind the formation of the American republic and the development of American democracy came almost exclusively from the British Isles and from France. These ideas and institutions expressive of the individual's right to think, to speak, and to develop freely in spiritual, economic, and political ways—to be a full individual without curtailment by priest or potentate—are the inheritance of English-speaking peoples over the globe. Nor is that inheritance horseless. The Thoroughbred, the Morgan, the ranch-adored Quarter Horse, and other lines of saddle horses now ascendant over America may be traced back to English stock brought to the Atlantic seaboard. Indeed, at the very time that Texas cowboy and Mexican mustang were at their zenith of blended action on long trails and fenceless ranges, an "American horse"—a horse from English stock—commanded zenith prices on those trails and ranges. He was already absorbing and supplanting the Spanish horse, just as Shorthorn and Whiteface were absorbing and supplanting Spanish longhorned cattle.

It was the Spanish, however, who brought the horses behind the Age of Horse Culture shared by Plains Indians, by English-speaking dispossessors of the Plains Indians, and by Spanish men on horseback. Blended with the fact of Spanish horses are three other facts: Spanish cattle, Spanish ranching on the open range, and the cowboy—a composite of Spanish, Mexican, and Anglo-American forms of life on horseback. All four of these facts enter into the history of *The Horse of the Americas*.

The subject is ample enough for the amplest of books. Segments of it have appeared in many places. Gallant, quick, and quixotic Robert B. Cunninghame Graham told an important part of it in *The Horses of the Conquest* (1930). Another part of it, disconnectedly put together, was issued by the Texas Folklore Society in 1940 under the title of *Mustangs and Cow Horses;* and this segment was materially added to by Walker D. Wyman's *The Wild Horse of the West* (1945). There are other titles that both delight and instruct, the best ones being perhaps more in the personal manner of Ross Santee, Will James, and Jack Thorp. Nobody wants to either read or write a book that, disregarding the highly important "art of omission," contains all there is to tell about Spanish horses in the Americas. Now, however, more nearly than any predecessor, Bob Denhardt has brought the story of these horses into focus and told it from the time that Columbus planted Spain's seed stock in the West Indies to the Palominos so conspicuous in Hollywood parades. Bob Denhardt knows horsemen as well as horses; he has made his own the traditions that live in the horseback world. These traditions he has absorbed in South America, where men who cherish both history and horses are organized to continue the *Criollo* (Spanish) breed, as well as in North America, where mere change is often mistaken for progress and the breed is rapidly dying out. He has, as almost every page of the present book attests, been searching for years through the records of Spanish-American history. Only such a combination of knowledge, training, and sympathies could have produced *The Horse of the Americas.* It is human as much as horse.

J. Frank Dobie

Preface to the Revised Edition

A quarter of a century has passed since the first edition of *The Horse of the Americas* appeared. In the intervening years both the publisher and I have received many requests for the book. None were available, and only seldom was a person able to obtain a copy by diligent search of the secondhand book market. Copies were rare and almost nonexistent. As a result we began discussing the possibility of a new, revised edition.

Several factors were considered before a decision was reached. Perhaps the key factor was that no other book had appeared that covered the same field. Some excellent books had been published that enhanced certain areas, such as Frank Roe's *The Indian and the Horse* and J. Frank Dobie's *The Mustangs*, but none covered the arrival, spread, and development of the Spanish horse in the Americas.

The last portion of the book, which covers the Western breeds developed in whole or in part from the Spanish horse, has been brought up to date in this edition.

ROBERT MOORMAN DENHARDT

Arbuckle, California
August 7, 1974

Preface to the First Edition

The following narrative is designed to tell something of the history of the Western horse and the part he has played in the development of the Americas. It is divided into several parts, which are in turn made up of relatively small and independent sections. Each of these represents an episode intended to illustrate some typical phase of the story.

Herbert Eugene Bolton, eminent historian, teacher, and friend, regrets that so many history books are merely dry compilations of historical facts. He feels that history should retain the vigor of its natural interest. His books represent an admirable fulfillment of this concept. In his famous round-table seminar this book had its inception.

J. Frank Dobie, the noted author, teacher, and master of Western folklore, insists that the beliefs and mores of a people are an essential part of their history. His widely read books incorporate this thesis, leaving no doubt of its effectiveness. Some of the most interesting parts of this book grew out of long sessions held in front of his hearth.

I have tried to follow the inspiration offered by these two scholars. There are other men who have offered valuable suggestions, particularly Lawrence Kinnaird, Paul Albert, and Juan José Fernández in North America, and Guilherme Echenique Filho, Roberto C. Dowdall, and Emilio Solanet in South America. The personnel of the Bancroft Library of the University of California was courteous and helpful, as were the other librarians in North and South America.

I wish to express my appreciation to the General Education Board, a Rockefeller Foundation, for the grant which, until the outbreak of war in December, 1941, assisted me in the early stages of the book.

Some of the information contained in this volume appeared under my signature in *Agricultural History, Anales de la Asociación Criadores de Criollo* (Argentina), *Anais da Associação dos Criadores de Cavallos Crioulos* (Brazil), *California Historical Society Quarterly, The Cattleman, Country Life,* Scribner's *Dictionary of American History, Hispanic-American Historical Review,* Texas Folklore Society's *Mustangs and Cow Horses, New Mexico Historical Review, Southwest Review, The Horse, Western Horseman, Westways,* and several other publications. They have courteously allowed portions of this material to be used in the present book.

ROBERT MOORMAN DENHARDT

Dallas, in the state of Texas
The sixteenth day of May, 1947

Contents

Illustrations

COLOR

BLACK AND WHITE

MAPS

The Horse of the Americas

Prologue

Thus moving ever westward around the globe, the horse had at last returned to the plains of America—a unique American Odyssey.

1. An American Odyssey

⸎§ *A Fountainhead*

It seems only right that America, apparently designed by nature to provide an ideal home for the horse, may be the fountainhead of equine existence. Only the fossils of the original horse which inhabited the Western Hemisphere remained when Columbus reintroduced Spanish-bred mounts into the New World in 1493.

One cannot help but wonder why the horse disappeared from the New World. Several plausible theories have been advanced concerning this extinction. Alfred Sherwood Romer places man in America when horses were still in existence. It is possible that prehistoric man, by continually hunting and chasing the herds, drove them from their best opportunities of livelihood. We know that early man hunted and ate horses in Europe, for their bones have been found in campfire remains, especially at Solutré, France. It is also possible that the increasing competition of the bison, which migrated to America, made living increasingly difficult on their chosen grasslands. Still another explanation could be the tsetse fly, which has been found in Pleistocene deposits. Less likely, but possible, is the theory that some epidemic or prolonged drought may have exterminated the animals. In any case, when the Spaniards arrived, there were no horses.

The ancestors of the horse can be traced back, from skeletal remains found in America, nearly to the beginning of the Tertiary period, without a single important break. During this long era,

6

Prehistoric horses. Facing page: top, *Eohippus*; below, *Mesohippus*. Above: top, *Parahippus*; below, *Merychippus*. (Courtesy University of California, from a film on the history of the horse prepared by the Extension Division of the university)

generally estimated at sixty million years, the horse, in its development, passed through astounding bodily changes. The most noticeable alterations, outside of size, took place in the teeth and feet. Horses developed on the open plains of a plateau region, where scanty, stunted herbage and probably short, succulent grass had been the rule.

The bodily changes which took place in the horse were so great that when one of the early ancestors was discovered a few years ago by the great paleontologist, Richard Owen, he failed to identify it and called his discovery the *Hyracotherium,* or "coneylike beast." Its relationship to the modern horse he did not even suspect. It was not until several intermediate skeletons were discovered that the modern scientist has recognized the true place of the *Hyracotherium* in the evolution of the horse.

Including Richard Owen's rabbitlike beast, a number of stages have been uncovered. Together these stages complete the cycle and trace the change from the first small, rodentlike beast to the modern equine animal. Each stage belongs to successive formations, and each stage is characteristic of its own particular geological environment. By far the most complete and best-known series comes from the Tertiary Bad Lands of the western United States.

With the exception of the *Hyracotherium,* the *Eohippus* is the oldest known horse. It was a tiny being no larger than a fox terrier, and it had a skull so small that it could be concealed in a man's hand. The eyes were directed laterally, and the orbital ring, or eye cavity in the skull, was not closed. The teeth were small and low-crowned, adapted for chewing the succulent plants upon which he fed. *Eohippus* had four toes on the front foot and three on the hind foot. The tips of his toes were tiny hoofs very much like those of later horses. All four of his toes reached the ground, but most of the weight was borne on the largest. The foot was still well padded.

Among the illustrations in this volume are two of these little horses on a sand bar adjacent to the area they usually grazed. They were not as fleet as later horses, but they avoided the large predaceous ground birds and their many other enemies by continual alertness and agility. *Eohippus* probably fed on the sandy and grassy areas of the lowlands, which bordered the underbrush of the densely forested regions. When frightened, he would scurry to cover much as the rabbit does today. Evidently his habits resembled those of some species of large rodents now living in South and Central America.

In the next stage, the horse was well advanced over *Eohippus,* although he retained many primitive features, such as the open orbital ring. The skull and brain increased in size. The teeth were larger, uncrowned, and still not adapted for grinding harsh, dry herbage. The outer toes were completely lost, and the weight of the animal was borne primarily by the center digit, though the side toes rested on the ground. This horse is called *Mesohippus.*

Remains of *Mesohippus* have been found most abundantly in the great Bad Lands of South Dakota and northeastern Colorado. Here the vast expanse was cut by winding streams and wooded valleys. Hackberries predominated among the vegetation, although willows and sycamores were plentiful. Evidently the bunch grass which was to be common in later geological epochs had not appeared. It is only reasonable to assume that these animals, still no larger than fair sized dogs, inhabited the stream borders where they fed on succulent plants for which their teeth were adapted.

The horse in the next stage is called *Parahippus.* In this step the skull more closely resembled that of the modern horse. The orbital ring was closed but not solidly fused, and the eyes, although still lateral, were now directed slightly forward. *Parahippus* had larger teeth than any of his predecessors. There was a noticeable increase in the length of his cannon bone. The central toe supported most of his weight, but his extra toes remained conspicuous even though they did not touch the ground. These developments, particularly the longer legs, indicate increased speed, an ability factor well calculated to meet the needs of the horse in his fight for survival in the plains country.

Even more like our modern horse was *Merychippus,* whose skull was broader than that of any earlier horse. The teeth had become higher and extended well into the jaws, an adaptation no doubt arising from the abrasive effects of the dry bunch grass and sand which had to be taken into the mouth. *Merychippus* is considered the typical three-toed horse, the side toes having receded and shifted farther back toward what is today termed the fetlock. These animals were comparable in size to our smallest ponies but entirely lacked their bulky thickness, being formed more gracefully.

Typical plains conditions were prevalent during the era of *Merychippus.* Hackberries, willows, cottonwoods, and sycamores bordered the streams, and many small lakes were formed in the broad river valleys. This plains country fostered an abundant animal life. Vast herds of mammals trod the prairies—huge mastodons, giant long-limbed camels, and small, delicate, prong-horned antelopes were among the most conspicuous. Thousands of these animals were trapped in the quicksand built up from the sediment deposited by the rivers, leaving a story in bone for the modern paleontologist to unravel.

Pliohippus is a direct predecessor of our modern horse; the skull was now larger and the orbital ring was solidly closed. The eye assumed a forward direction. Only certain depressed areas on the side of the skull showed the relation to the open orbital ring which occurred in earlier forms. Long, curved teeth extended three inches up into the head, and the side toes now were only sliverlike splints.

Last in the ancestral series is *Equus,* the horse of the Glacial epoch. Every feature of skull and skeletal specimens of the *Equus* has its counterpart in the skulls of certain types of present-day horses, whose technical name is also prefixed with the word *"equus."*

Although one may quickly outline on paper the development

from *Eohippus* to *Equus,* actually these specimens are separated by a period of approximately sixty million years. During these long ages the early horse was extremely abundant, and at least four migrations took place from the New World to the Old. Of these migrations, only the last group of wanderers survived. In the Americas the original stock died out completely. From Manitoba to Patagonia the horse disappeared from the Western Hemisphere.

With the advent of the Spanish *conquistadores,* the American horse, after wandering far over the earth, journeyed again to what may well have been his homeland and to his rightful place in the history of the hemisphere. Thus moving ever westward around the globe, the horse had at last returned to the plains of America—a unique American Odyssey.

Spanish Background

Perhaps the greatest Islamic contribution came in the form of horseflesh and horsemanship. The Spaniards obtained not only Moorish horses but also the Moorish manner of riding. It is small wonder that Spain became a horseman's country where even the word "gentleman," caballero, *meant "horseman."*

2. . . . jineta en ambas sillas

The Coming of the Moors

United by the Mohammedan religion, the Arabs began their ruthless conquest in the year 634, a conquest and domination which was to cover much of Asia, Africa, and Europe. In the first years of their expansion they entered into lands where the people, like themselves, were horse-lovers and horsemen. This was especially true in Syria, Persia, Egypt, and North Africa. An exchange of horses was the natural outcome. By 696 most of northern Africa was conquered, and the capture of Ceuta gave the Arabs the necessary base for invading Spain and Europe from the southwest.

At first a small detachment of Moors entered Spain in the year 710 under the command of General Muza. During the following year a powerful army of some seven thousand Moslems crossed the Strait of Gibraltar, and a little later five thousand more cavalry joined them. Most of the cavalry were Berbers, and it may be presumed that most of them rode Barb horses. Some two hundred were Arabs; these may well have had some Arabian horses.

The first battle that the Moors fought against the Europeans, who were under the command of Roderick, was the battle of Guadeleta. It lasted three days. According to Dhobi, the Moorish historian, the defeat of the European army was due to inadequate cavalry. By this he meant two things: first, they were too few; and secondly, they were too cumbersome.

The agile Moorish horsemen and their beloved steeds found

little competition offered by the awkward knights of medieval Europe. Spain was conquered by a new cavalry, differing from the European both in horse and in horsemanship. The Spanish horses, especially those of Andalusia which had been so famous in former days and celebrated by the poets, had become decadent, and were apparently awaiting the Arabic invasion to rise once more to the heights of fame.

The Spaniards, in spite of their well-intended objections, were being brought a more advanced culture which included a new tradition of horsemanship and the best equine blood to be found in the world at that time. The Moslem invaders were no amateur horsemen; nor were their horses common plugs. Spain was benefiting from a heritage that came from Arabia through Syria, Egypt, Nubia, Zeneta, and Barbary.

It can be understood why the invaders of Spain, with these great equestrian nations a part of their background, were exceptional horsemen, and why the blood of their horses proved so excellent. Each and every man was notable in his own right as a horseman. All of them had always lived on, and for, horses. Their lives had been dedicated to warfare on horseback, and they took pride in the beauty of their horses. They had developed and kept small, light, fast horses, directly contrasting with the large, heavy, slow horses utilized in Europe by the medieval armored knight.

In North Africa, whence the Arabs and Moors entered Spain, good horses had been bred for a period of at least sixty years. The blood had been improved principally by the entry of animals from Egypt, Syria, and Arabia. These horses were of Asiatic or Oriental type with a straight or concave profile. Actually, the type that is considered today the true Arab probably came only with the higher class of chiefs. The large majority of horses brought to that land, however, did have good Asiatic blood in their veins, a blood which consistently produced small, swift, and hardy horses.

The Arabic domination of Spain was completed by Syrians and

Berbers for the most part. During the first forty-five years after the entry in 711, the Moslems ruled Spain through chiefs who were called "emirs." The Emir was subordinate to the Guali in Africa who was in turn subordinate to the Caliph of Damascus. In 756 the Emir in Spain declared his independence.

Spain may have submitted to the forceful entry of the Arabs, but rebellion simmered underneath. For a period of seven hundred years—in fact, until the fall of Granada and the discovery of the New World—she never relaxed her efforts to resist her new master. In the course of this intimate contact, it was only natural that each should learn from the other. Perhaps the greatest Islamic contribution after agricultural reforms came in horseflesh and horsemanship. The Spaniards obtained not only Moorish horses but also the Moorish manner of riding. During the long domination of Spain by the Moors, even during truces, natives continually raided enemy camps. As these excursions were made on horseback and as the Moorish horses were one of the primary motives for attack, it is small wonder that Spain became a horseman's country where even the word "gentleman," *caballero,* meant "horseman."

The horses of the Berbers, generally called "Barbs," were introduced into Spain in the largest numbers and multiplied mainly in the middle and southern part of the Peninsula, where the climate, the fertile valleys, and the succulent pastures invited increase. Horse breeding had been neglected here before the invasion, but with the arrival of the Moors it once more became important. So successful was the activity that soon the horse of this area became world famous and retained that fame for four hundred years. It held the same place in the horse world between 1200 and 1600 that the English Thoroughbred occupied after 1700. Spanish horses were taken to England during the time of William the Conqueror (1027–87) to improve the heavy race of Norman horses already there and played a most important role in the establishment of the modern Thoroughbred.

At the time of the invasion, Spain, in common with the rest of Europe, rode *a la brida,* which was the style of chivalry and medieval knighthood. This type of riding was distinguished principally by the use of long stirrups. The knight sat on his horse like a bareback rider, legs hanging straight down. Normally, the horse being ridden *a la brida,* as well as the rider, was covered with armor. The saddle was made to enclose partially the hips of the rider and so support him during battle. Thus the heavy horse was desired in northern Europe because the armor on horse and rider made a sturdy horse essential, hence the animals were bred for size. Since they were to carry weight, speed and agility had to be sacrificed. This breeding for size contributed to the great draft breeds that arose in the area around the English Channel, but it was a disadvantage to the Spaniards in the year 711.

The new school of riding, *a la jineta,* brought into Spain by the Moors and adopted by that country, was much more effective for battle than the old method. For riding *a la jineta,* the saddle was placed well up on the withers, and because the stirrups were considerably shorter than for the *a la brida* fashion, the rider appeared to be almost kneeling. The Moorish saddle was so shaped that when the horseman had his feet in the short stirrups, he could not easily be dislodged, as his knees were brought up under a swell in the fore bow. A cavalryman riding *a la jineta* had more freedom to move in the saddle when fighting, and thus the Moors easily rode around the awkward, heavily armored Spanish knights riding *a la brida* and speared them at leisure. After observing this advantage, the Spaniards soon discarded the *a la brida* saddle and began designing their own *a la jineta* saddle. It was small, light, and fitted to the horse, and so made riding easier than the old heavy and cumbersome one. The front tree of the new saddle was always higher than the rear, a characteristic still found in the western saddle, which is an outgrowth of the Spanish-built Moorish-type *a la jineta* saddle brought to the New World by the *conquistadores.*

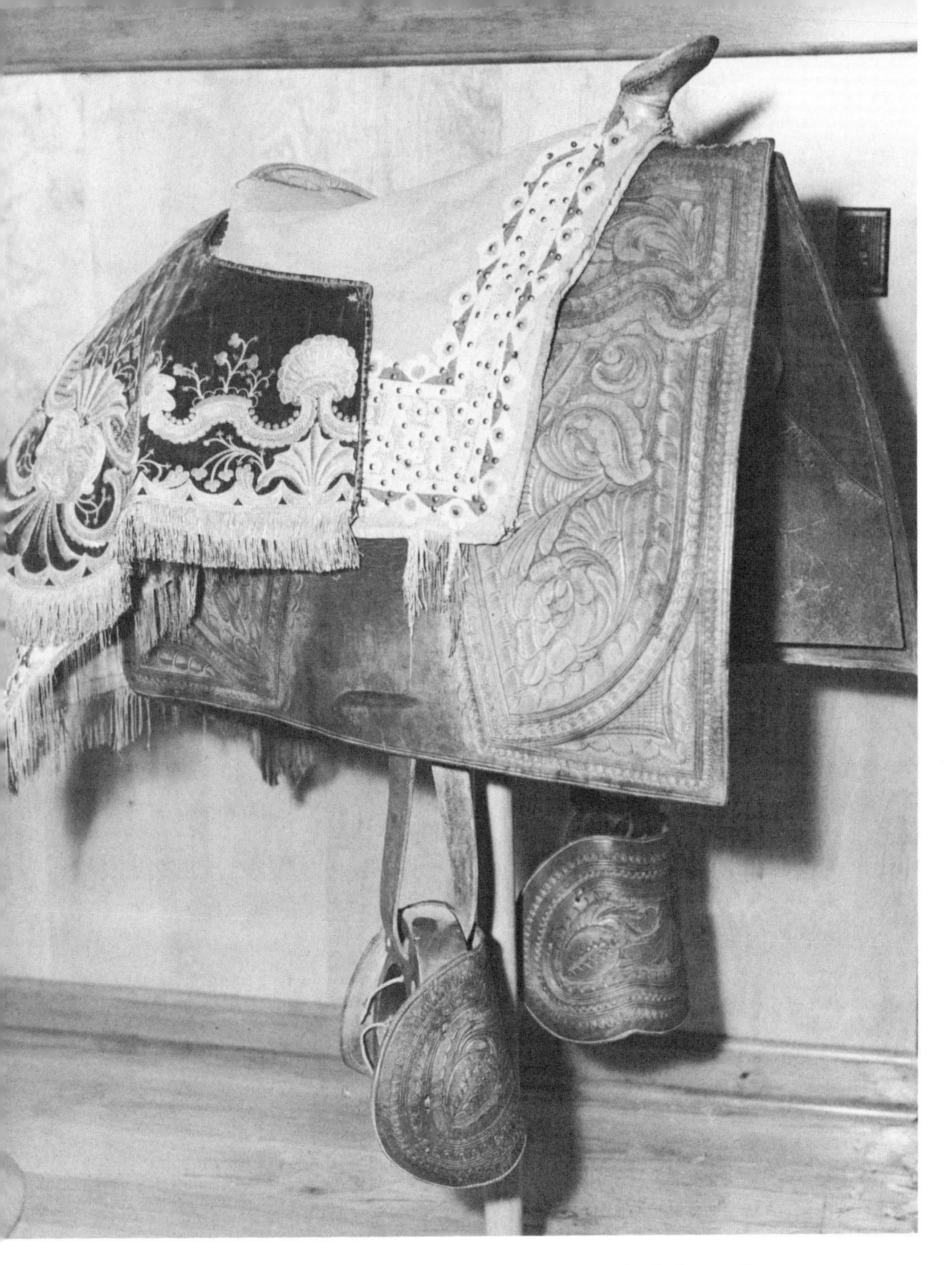

A type of saddle used in Spain many years ago, particularly on fiesta days. The stirrups are hand-carved of wood. The covering back of the cantle is of cloth, with silver thread making the designs. (Photograph by Lucille Stewart, Los Angeles)

During the long Moorish domination and fierce fighting which preceded the final expulsion, the Spaniards came to master both styles of riding. So much importance did they attach to the ability to use both methods that the highest praise a gentleman could be given was to say he rode well "in both saddles," which meant that he was able to ride equally well either in the *a la jineta* or the *a la brida* style. This fact was often inscribed on a hidalgo's tombstone.

When the Spaniards had become proficient in this manner of riding and had obtained the light Moorish horses, they were on a more equal footing with their opponents and the Moslems were soon driven out of Spain, almost at the same time as the discovery of the New World. Consequently, when the Spaniards started their long movement across the Atlantic, they took with them their new horses and the new style of riding, both of which were the best in the world at the time. All the *conquistadores* rode *a la jineta* and were proud of it. The Inca Garcilaso de la Vega proudly says in his chronicle of the conquest of Peru, *"Mi tierra se gano a la jineta,"* "My land was conquered *a la jineta.*"

Jineta, Gineta, Genete, or Zenete

The word *jinete* comes from the Arabic, or more properly, the Arabianized Berber word, *zenete.* Some dictionaries give the derivation as from the Greek word *gymnetes,* but the eminent Spanish and Argentine scholar Angel Cabrera points out the falseness of this claim in an article on the *jinetes* in the official publication of the *Criollo* Horse Breeders in Buenos Aires. The Zenetes were a powerful tribe or nation in the north of Africa, famous for their ability as horsemen and as warriors. They took a very active part in the warfare between the various Moorish dynasties and constituted the bulk of the Mohammedan cavalry that invaded Spain.

Those who have read the Castilian classics of three hundred years ago know that the word *jinete* did not then signify, as it does in much of the Spanish-speaking world today, someone who rode on horseback. This usage is modern. Three hundred years ago everyone who rode, even if he were mounted on a burro, was called *caballero. Jinete, ginete, genete,* or *zenete,* as it was variously used by the earlier writers, had a pagan connotation, since it referred to the Moorish tribe of North Africa from which it was derived. Thus in the *Crónica de Alfonso el Sabio* the Saracen cavalry is described as being made up of 500,000 men, part from Andalusia and part *"Genetes,"* meaning men from North Africa. The Padre Mariana in his *Historia de España* tells how John I of Castile was killed when he was thrown from his horse in the presence of some *"jinetes"* recently come from Africa. Alfonso III of León copied the *"Zenete"* cavalry of North Africa by forming a similar branch of light cavalry which became known as *Lanzas Zenetas.* As time went on the word *zenete* or *jinete* was applied to all horsemen who rode in the fashion of North Africa or the Moors. Since the Moors were excellent horsemen, it also came to be applied to a person who was a skilled rider—he would be called *muy jinete.*

The horsemanship of the Zenete people became most popular, and their riding forms are still used today in North Africa from Morocco to Tunisia. Spain first accepted *a la jineta* as an additional method to be learned along with *a la brida.* Gradually the new manner replaced *a la brida* for normal riding. Even Sancho Panza in *Don Quixote,* when he became governor of Barataria, received the honor mounted *a la jineta* on a mule. Only *a la jineta* riding was practical for many of the favorite horseback sports such as *juegos de cañas,* literally "caneplay." This was an ancient Arab game brought to Spain by the Moslem conquest. It was, of course, played on horseback and mocked a battle, the cane sticks taking the place of javelins. The players threw the sticks at one another, and the point of the game was to catch them without being hit. In warfare, everyone recognized the superiority of *a la jineta* riding after the conquest of Spain.

Riding *a la jineta*. (From *El Libro de la Monteria*, by Gonzalo Argote de Molina [Seville, 1582])

Riding *a la brida.* (From *Handbuch der Waffenkunde*, by Wendelin Boeheim [Leipzig, 1890])

A la jineta riding seems to find its ideal location in the wide-open spaces and in hot countries, where the simplest armor and tough, agile horses with good mouths and great hearts are needed. This style of riding flourished in the New World.

Captain Bernardo de Vargas Machuca, author of a most interesting book called *Libro de Exercicios de la Gineta,* which was published in Madrid in 1600, says that the art of riding *a la jineta* came to Spain from the Berbers of North Africa, and from Spain it was taken to the West Indies where it was perfected more than in any other country. This indicates that he considered the *vaqueros* and gauchos equal to the Arabs and the Moors in riding.

In later years, when Spain came under the influence of Austria, once again the old equestrian art of *a la brida* regained popularity. Some writers felt that soon the manner of riding *a la jineta* might be lost in Europe. Luis Banuelos y de la Cerda in the prologue of his book *Libro de la Gineta* stated that in the future one would have to go to the New World for masters of equitation *a la jineta,* since this system was being so badly neglected in Spain at the time. In New Spain, or Mexico, there were past masters riding *a la jineta.*

Today, although the old words *jinete, jenetear,* and *jeneteada* are no longer encountered in the Spanish-American vocabulary as they used to be a few centuries ago, *a la jineta* riding, with its short stirrups and neck reining, is found from the plains of Alberta in Canada to the Patagonian plains of Argentina. In some countries, such as Chile, there is used even now a bit and a horsemanship almost identical with those of the Berbers of Africa. Today throughout the Americas (except for an unbroken horse) the stirrups have been lengthened for comfort. After all, it is no longer necessary to rise in the stirrups to use a lance, or is there danger of being knocked out of the saddle if one cannot dodge a lance. In the old days in Spain and the New World, when the stirrups were lengthened, the term *a la bastarda* was applied, a name whose origin is obvious.

◆§ *Arab or Barb?*

When speaking of the horses introduced into Spain by the Moorish invasions, many writers, depending upon their knowledge and their feelings, have said that they were Arabs or of Arabian type. It is entirely possible (although I know of no historical proof) that some of the top-ranking leaders of the Moorish invasion did have horses imported from Arabia, or rode horses which might loosely be called purebred Arabs. Nevertheless, most of the horses taken were Barbs. Barbs are not Arabs or Arabian type any more than the Thoroughbred is Arab or Arabian type, although both have Arabic blood. The contemporary chroniclers, historians, and poets speak only of African horses, or Moorish horses, or Barbary horses, or Tlemcen (near Oran) horses, and not of Arabian horses.

The historian Ibn Kallikan claims that Tarik took with him into Spain twelve thousand horsemen of which only twelve cavalrymen were Arabians—these twelve might conceivably have had Arabian horses. The remaining men were Berbers, Zenetes, and other North Africans. Does it seems probable that 11,988 Arab horses were brought all the way from Arabia for the invasion or that so many purebred Arabs were found in Morocco for the troops? This is hardly likely when there were already available for these troops their own excellent Barb horses, of which historians tell us they were very proud. The mistake lies in assuming that the Mohammedan wave of conquest carried along with it, from the deserts of Arabia, Arabian horses. The Moslem conquest can be likened to a wave on the ocean which does not carry the same water with it.

Another indication that the best imported horses at this time were Barbs is the fact that the most famous and celebrated line of *jineta* horses that developed in Spain, those called *Guzmanes* or *Valenzuelas,* was founded by a Barb stallion, Guzmán, who was taken there by an official of the King of Morocco. In the appendix to his French translation of *El ornato de los Caballeros y el atributo de los valientes* (which was written in Granada in the fourteenth century by Ali Ben Abderrahman Ben Hoderd), Luis Mercier—one of the few historians to make the fact plain—clearly shows that Barb and not Arab horses were the foundation stock brought by the invasion.

If no other indications were present, one might even guess that the early horses were Barbs from the many contemporary illustrations that exist. In paintings and murals, the horses shown all have a tail set low in a sloping rump in the Barb fashion, while one of the outstanding features of the Arab horse is his high, level croup and his distinctive tail set. Surely some of the artists would have depicted this important feature if it were present. Likewise, the compact shortness of the Arab is not always shown. The illustrations clearly indicate other interesting facts, such as the different methods of riding and the different equipment used. In most of the pictures both Moor and Christian are seen riding the same type of horse, as they undoubtedly were, after the invasion had brought the lighter horses to Spain.

The King and Columbus both speak of the cavalry which Columbus took with him on his second and third voyages as *jinetes*. These soldiers were cavalry schooled in riding in the *a la jineta* manner, and it must be presumed that they had horses proper for their branch of the expedition. Such horses were called *jacas jinetas* and were bred in Andalusia especially for *jineta* riders. Very likely the modern Hackney pony derived its name indirectly by way of France from *jaca,* which meant a well-trained, light saddle animal.

The first horses were gathered in southern Spain from the provinces of Córdoba and Andalusia. These horses were the outgrowth of some seven hundred years of crosses of Barb with the native horse. It is possible that there was also an occasional sprinkling of Arabian blood. These animals no longer showed the large, thick body and coarse, hairy legs which indicate north European and Germanic origin; they were light, clean-legged saddle horses, not the plate-armor type. So many Barb horses had been brought in and crossed on them that they were more like the Barb than any other horse. Although they were light animals, they were broad between the shoulders and across the hips and deep through the heart—characteristics that the Barb often lacked. In other points they were like the Barb, being of small to average height, having on occasion a long, round body, a strong head sometimes Roman-nosed, thick-necked, with full loin, sloping rump, the color often some shade of line-backed buckskin, and a smooth gait ideal for a saddle horse.

The best-known horses, small and sturdy, came from a mountainous region in southern Spain called Ronda, which formed part of Andalusia in the province of Málaga in the old Kingdom of Seville. It may be just coincidence that these horses close to the ports of embarkation—for southern Spain was the origin of the early expeditions to the New World—had so many of the characteristics of the present Western horses, especially the *Criollo* of South America and the Quarter Horse of North America; but would it not seem more of a coincidence if the latter did not show their heritage?

☙ *"Caballeros" Should Ride Horses*

The transmission of livestock, particularly horses, to the New World was difficult for Spain at the beginning of the sixteenth century for several reasons. Spain was not so well stocked that she could go on indefinitely supplying two continents, both bigger than Europe, with horses. In fact, even before 1492 she realized the need of encouraging the breeding of cavalry horses. An ancient law, with an eye to horse breeding, decreed that *caballeros* should ride on horses as "honor and tradition" demanded. Despite this law, during Columbus' time the use of mules had become increasingly popular. The Catholic kings had tried vainly to stop this practice, which was considered effeminate, especially since the number of horses available for the royal cavalry had decreased to an alarming extent. The King had too many difficulties and plans to allow this important branch of his army to become inadequate. This accounts, too, for the law restricting the export of horses and mares without a special license. At the same time the Spanish king continued to punish all those who ignored the old laws commanding gentlemen to ride horses and not mules. It is interesting to see that Columbus was given special dispensation to ride a mule by the King. The letter, dated the twenty-third of February, 1505, granting him this privilege, said that because of the many hardships he had endured, his health would not permit him to ride a horse. This may seem strange today—that it should be easier to ride a mule—but it must be remembered that at that time the horses were not gelded, and consequently being mounted on a spirited stallion was entirely different from riding on a mule.

The upshot of the affair was that the Emperor issued an elaborate decree obliging the *caballeros* to observe the old statute that all gentlemen should ride horses, threatening to destroy the mules if they continued to ride them, and forbidding entirely the export of horses from Castile. It seems probable that at first this decree was observed rather strictly. In fact, certain mules paid the penalty (for the sake of justice, one may presume) in Valladolid and other towns. However, when so many horses were lost in the disastrous expedition against Algiers in 1541, in the following year the Emperor had to grant numerous exceptions from the law.

Coaches and litters were becoming popular at this date, and they also tended to decrease horseback riding. Fifty years later, in 1634, Prudencio de Sandoval wrote in his book on Charles V that the men of the realm had become like feeble women (riding in carriages and on mules) and it was to be feared that the day of judgment God had threatened was near at hand, when the leaders of the nations should become effeminate as they were in the last days of the mighty Goths, when Spain was lost to the infidels. Thus a combination of factors resulted in an inadequate supply of horseflesh in Spain, which in turn had its effect on the number of horses shipped to the New World. Nevertheless, enough were available to supply a foundation for the American horse in the years immediately following 1492.

The Advent

Veinte y tres de Mayo de 1493 Archivo de Indias. El Rey e la Reina: . . . nos mandamos hacer cierta armada para inviar á las islas e tierra firma que agora nuevamente se han descubierto . . . y . . . veinte lanzas jinetas a caballo: . . . é los cinco de ellos lleven dobladuras é las dobladuras que llevaren sean yeguas . . . . .

3. . . . and the Horses Multiplied

◆§ *Columbus Also Saw Mermaids*

"Horses are the most necessary things in the new country because they frighten the enemy most, and after God, to them belongs the victory." Pedro de Castañeda de Nagera voiced the opinion of all the conquerors when he made this statement, for every chronicler of the period reiterates the sentiment. Obviously the great conquest would have been impossible without horses. The Spaniards, having slight idea where they were going and less what awaited them, had to rely almost wholly on their faithful steeds. Most of the *conquistadores* had slender means, scanty forces, and little assistance from their king. When one realizes the tremendous trip they were forced to make from far-off Spain in the bobbing cockleshells they called ships, one cannot but admire their enormous courage and fanatical faith in their own destiny. However great was their ability, it was exceeded only by their ignorance, but therein lay their power. Had they known what lay ahead, well might they have trembled, but on they strode, clad in their puny armor, the Christian faith, and a blissful delusion.

One of the more romantic features of the conquest was the use of the horses by the *conquistadores.* Indispensable as they were to these men, they furnish us with a group of stories that will never grow old as long as red-blooded adventure and heroism appeal to human nature; tales such as the one about Morzillo, the horse who had the strangest destiny of his kind, the incredible ride of Gonzalo

Silvestre through the Everglades of Florida, or the feats of Motilla, the renowned steed of Sandoval, who might rival the horse of *El Cid Campeador*. These and many more stories are found just as the *conquistadores* have recorded them in the early chronicles of the conquest.

Herrera, Oviedo, Navarrete, Acosta, Garcilaso de la Vega, and Zárate, all famous chroniclers of the period, throw light on the horses of the early Spaniards and recount many interesting stories. Columbus' first brilliant military success in the New World occurred in Hispaniola in 1495. When he saw that a large group of natives were arming, he decided that he should strike first. He took to the field with two hundred foot soldiers, twenty horses, and a number of dogs. The effect of the horses upon the Indians, who saw them for the first time, was startling, and they were badly defeated. It was undoubtedly this first combat that made Columbus realize the vital importance of his horses and that caused him thereafter to demand that the King send over horses with every ship leaving for the New World.

Some time later, while sailing, Columbus noticed that the horses were becoming extremely poor from standing day after day in the boiling sun on the narrow decks. Consequently, instead of proceeding directly to Puerto de Gracia as planned, he put about to where a great river fell into the sea and ordered the men and horses ashore for a rest. There is an old saying that goes "It's a wise commander who looks after his men's feet."

The Spaniards did not know for a number of years that there were no horses in the New World. In fact, until some time after Columbus' death they thought that there were. Columbus himself on his fourth voyage wrote the King: "It was told that those on the shore of Veragua [Panama] had horses which they used in battle." But then Columbus also saw mermaids on several occasions.

The Twenty-third of May

During the earliest years of exploration, horses were sent in every ship leaving the Spanish ports. Each of the conquerors, when he entered into an agreement with Charles V and his son Philip II, was bound to take a certain quota of stallions and mares. Columbus himself never brought more than one hundred horses in all of his voyages. The ones which he carried must have been generally like those painted by Velázquez and ridden by Phillip IV. They were, as Robert Cunninghame Graham describes them in his *Horses of the Conquest,* short-backed, without much daylight showing beneath their bellies, and admirably suited for the hard work of the conquest. Their adequate pasterns made them comfortable to ride, while their legs, not too long and firmly jointed, showed that they were sure on their feet. Their descendants in the Americas lost flesh and gained sinews and angles during succeeding generations, while their looks were in part replaced by toughness. They developed until the Western horse became a race capable of unbelievable feats of endurance.

There is a tradition generally found in English histories that Columbus during his lifetime thought only of discovering India and that the Spaniards gave consideration solely to gold and glory. At least one other purpose, colonization, is clearly expressed in Columbus' report to Luis de Santangel, who controlled the finances for Castile. In this report we find Columbus telling that the newly discovered land, with its mountains, hills, plains, fields, and soil, was excellent for planting and sowing and for the breeding of livestock. In these words we find a hint of another motive behind Spanish occupation. In the Old World, Spain had two great industries, mining and stock raising; it is hardly surprising that she should turn to the same pursuits in the new lands.

When Columbus was weighing anchor at Cádiz at the beginning of his second voyage of discovery and conquest of the New

World, he had with him, among numerous other items necessary for the colonization of new land such as Hispaniola, five brood mares and twenty stallions. The royal cedula is still preserved in the archives of the Indies, as are most of the original correspondence and documents relating to the New World. It reads as follows:

> The Twenty-Third of May, 1493. Archive of the Indies. The King and the Queen: Fernándo Zarpa, our Secretary. We command that certain vessels be prepared to send to the Islands and to the mainland which has been newly discovered in the ocean sea in that part of the Indies, and to prepare these vessels for the Admiral Don Christopher Columbus . . . and among the other people we are commanding to go in these vessels there will be sent twenty lancers with horses . . . and five of them shall take two horses each, and these two horses which they take shall be mares.

The above is probably our only record of the first horses to be sent to the New World. In it we see that, although the horses were to be used for warfare, there was also clear provision made for propagation. Fifteen of the cavalrymen or lancers had stallions and the other five each had two mares. This made a total of ten mares and fifteen stallions, twenty-five horses in all, which were to be the first horses to set hoof in the New World.

Because of the Spanish practice of riding stallions, the only urgent measure concerning horse breeding was to be sure that mares were occasionally brought to the New World. Stallions would arrive with every contingent of soldiers. Columbus well knew this, and we find him stressing in his letters to the King the need of including some mares in each vessel sent to the Indies. In January, 1494, he wrote to the King and Queen, through Antonio Torres, as follows: " . . . each time there is sent here any type of boat there should be included some . . . brood mares."

The actual lists of animals that were sent to the New World once the occupation was under way are difficult to find. On April 9, 1495, there were dispatched four caravels with a quantity of livestock aboard, including six mares. On April 23, 1497, fourteen mares

The method of loading and transporting horses on boats in the sixteenth century. (From *Manejo real en que se propone lo que deben saber los cavalleros,* by Manuel Álvarez Ossorio y Vega [Madrid, 1769])

were sent. Because difficulties arose in exportation, we have a record of some merchants who obtained special permission to ship 106 mares from Seville, San Lúcar, and Huelva. These merchants were the same who provided the horses for the twenty lancers on Columbus' second voyage. These original lancers also got into a little trouble since they apparently sold their good horses and embarked with just common nags. Columbus was very much upset and wrote to the King about the situation in January of 1494.

In 1498, when the Admiral was organizing his third voyage, he was authorized to carry to the island of Hispaniola forty *jinetes* (horsemen) and their horses. In 1501 Don Nicolás de Ovando took only eighteen horses, but they were the best available *("de distinguida casta")*. According to the records, by 1500 the Crown had one ranch on Hispaniola that boasted sixty brood mares. In other words, in less than ten years large breeding ranches were operating.

A few more horses were undoubtedly sent from time to time, although the records are either lost or have not been uncovered. It is certain, however, that because of the European interests Spain had at that time, she could not allow too many horses to leave for America; therefore, we may presume that after the initial stock had arrived, very few additional animals were sent.

Into the hastily assembled ships of Columbus' second expedition were crowded as many horses and cattle as was practical; and upon his arrival in Hispaniola, his first act was to establish ranches, for he had included stockmen among his colonists. The importance that Columbus attached to these early ranchers is evidenced by one of his letters to the King in 1496, in which he said that the men thirsted for gold so much that it seemed to him that licenses should be granted to allow them to search for gold at certain times of the year in order to encourage further agricultural pursuits.

As has been indicated, there are very few records of early Spanish shipments of horses to the New World, and even fewer that contain numbers or descriptions. It was only when unusual circumstances arose that this information was recorded. The normal traffic that left for the New World during the time was too commonplace to be recorded for the benefit of future readers. Generally, there was merely this simple statement attached to the end of a shipping list, "and stallions and mares" *("y caballos y yeguas")*. In a cedula, or directive, to Juan de Fonseca from the King, dated April 9, 1495, we find one of the exceptions. Fonseca was to tell Diego Carillo to take to Columbus certain cattle, horses, mules, sheep, goats, and hogs. Whenever shipments of livestock arrived, every effort was made by Columbus to facilitate breeding, and by 1501 amazing progress had been made in the New World.

Despite certain regulations set up later to prohibit the export of horses from Spain, they were sent on occasion with the permission of the authorities. On December 23, 1507, the King ordered the liberation of 106 mares destined for the West Indies and held up by the officials of the *Casa de Contratación,* or House of Trade. We should never have known of this consignment had it, too, not had difficulty in Seville. Doubtless many horses were shipped with the House's permission without an express order from the King. For a time the King provided, at Columbus' request, free transportation for brood mares to Hispaniola.

The voyage across the ocean sea *(Mar Oceano* it was called) from Spain was a trying ordeal for the horses. It was a long way to the new colonies. In the sixteenth century a good passage from Spain to the New World was from two to three months, and the loss of animals sometimes exceeded 50 per cent. The boats and our modern vessels have about as much in common as trucks and tricycles. The early ships were high-pooped, low-waisted caravels with many little cannon, their tops hanging on the masts like miniature forts. In order that the roll of the ships would not throw the horses, the animals were suspended in hammocks stretched under their chests. Here, suspended between the beams and the underdeck, they hung, day and night, with only the monotonous wash of the water in the bilge to lend variety. On the smaller craft the horses stood on the main deck, side-lined or hobbled in fair weather, and tied down during storms. There they remained uncovered, rain or shine, encouraged only by an occasional friendly word or slap from their masters. In a calm, when the ship lay like a log close to the equator for weeks, and water failed, the sailors had to throw the horses, for which they could no longer spare water, into the sea. Hence the phrase, "the horse latitudes," so frequent in all books of voyages of the sixteenth and seventeenth centuries. This transatlantic voyage of about four thousand nautical miles was a trial worthy of the intrepid men and their faithful horses, and one through which they passed as only the brave can.

His Name Became Synonymous with God

Once in their new home, horses spread rapidly. José de Acosta, in his history written at the close of the sixteenth century, was impressed by their increase. He says that the horses "multiplied in the Indies and became most excellent, in some places being even as good as the best in Spain, good not only for fast messenger work, but also for war and the parade." He also tells of the many mules used for overland commerce and notes that there were few asses, as they were not needed.

The rapidity with which the horses spread in the New World is more clearly understood when one remembers that they were running wild from Tucumán to the borders of Canada by the close of the seventeenth century. The appearance of horses in the new continent would have influenced the cultural development of the American Indians sooner had it not been for the Spanish efforts to prohibit their use by the natives. On the frontiers, however, they were gradually obtained and meant not only an increased food supply but also greater resistance to the whites. As Roy Nash says, aside from Englishmen out for exercise, there are few people in the world who will walk when offered a ride. The pedestrian Indians, who had always done their best to steal up on their prey, gazed on the Spanish horseman in wonder and, as soon as they saw that the strange creature was not all one animal, did not rest until they owned horses. They were not long in discovering that a stolen horse would carry an Indian as well as a Spaniard, and whole tribes became equestrian.

The Indian's enthusiasm for the horse was as great as the present-day enthusiasm for automobiles. He ate horses; he drank melted horse fat; he shampooed his head in the blood of the horse with the idea of giving himself more strength. He twisted horsehair into rope; the hide he made into his couch, his clothing, his tent, his saddle, and his shoes. The skin of the foreleg he transferred to his own leg as a puttee, and the skin of the hock he made into a boot. And, when an Indian died, his horses were decked with bells, glass beads, and feathers, led solemnly in procession, and often staked to the dead man's grave or sacrificed. Horsemanship became the epitome of all manly virtues, an art to be learned as soon as a child could walk. Stirrups were not in general use on the Indian frontiers. Standing on the right side, one hand holding the mane and the rein, the other grasping a wooden lance, the Indian vaulted on his horse at a bound, and then woe to his enemy, man or beast. It is small wonder that horses were worshiped and their name became in most tribes synonymous with God.

It was not long until horses became so numerous that they were hunted instead of bred. Horses were introduced into the pampas of South America in 1535 by Pedro Mendoza. When forced to abandon the early settlement where Buenos Aires now stands, he reportedly turned loose five mares and seven horses, although it seems more probable that they would have been salted down for provisions (see page 163). Guzmán states that the progeny of these horses so increased that by 1600 they could not be counted but appeared in great droves and penetrated even into the mountains.

There is little doubt that wild horses permeated the country, but as for their origin, it is more likely that trade was established across the Andes and that the horses of Tucumán came not from Mendoza, but from Chile. However this may be, the Indians soon lost their fear of the animals, and fifty years after the date that Mendoza abandoned his horses, the Patagonians had become horse Indians and hardly stirred a step on foot.

Félix de Azara gives an interesting account of the amazing number of wild horses on the pampas at this time: "In the beginning horses were so scarce that Domingo Martínez de Irala, the governor of Paraguay, bought in the year 1551, in Paraguay from Alonso Pareja, a black horse with a white blaze on his forehead, and a white stocking on his near forefoot, for four thousand gold crowns, to be paid out of the fruits of his conquest." In the year of Irala's death, 1557, he left twenty-four horses. Azara adds, "Horses abound so much today [1800] that the poorest Gaucho owns horses, and they are worth about two dollars apiece." At the second founding of Buenos Aires in 1580, Juan Garay found that the whole province was full of wild horses. The herds flooded the entire pampas, from the shores of the Río de la Plata to the Río Negro, and were even found in large numbers in Patagonia. Father Bernabé Cobo, S. J., says that in Rio Grande do Sul a horse was worth nothing. To illustrate how numerous horses became, there was a common saying, "In Montevideo the beggars ride."

A Frisky Mare

It was not only in South America that the horses multiplied rapidly. Gonzalo Fernández de Oviedo y Valdés wrote (1570): "There were no horses found on Española [Hispaniola] or on any other island, and from Spain were brought the first mares. There are so many now it is no longer necessary to bring any more over. The Island of Española has produced so many horses that they have been shipped to other islands, that were inhabited by the Christians, and now there is an abundance there also." Because they had increased so rapidly, the price of a broken colt or mare soon dropped to four or five *pesos de oro,* or even less. This low price did not continue, as the demand increased and the supply was severely taxed.

After 1510 the surge of expeditions to the mainland drained the supply of horses on the islands to such an extent that it was feared they were going to be left barren. The resulting jump in price caused by this sudden demand was better than 100 per cent, for we find horses bringing two hundred pesos in 1530 and five hundred pesos in 1538. This demand created a boom for the island breeders. The *conquistadores* demanded island-bred horses, not only because they were easier and quicker to get, but also because they stood the climate better than the European-bred animals. Cuba, Jamaica, and Hispaniola became great rival centers, competing in their efforts to gain the favor of the prospective *"conquistadores"* and to sell them supplies. When the great expeditions, such as De Soto's from Cuba, Garay's from Jamaica, and Heredia's from Hispaniola, sailed from the islands to the mainland, they left in the hands of the island breeders thousands of *pesos de oro*. The colonists of the new land who had settled down to stock raising prospered. It is interesting to note how many of the three hundred original colonists of Columbus became wealthy by raising stock. More prospered from the soil than ever discovered gold. Dried meat and horses were a prerequisite to every expedition. The names of most of these ranchers have been obscured by time. However, one became so enthusiastic about ranching on Hispaniola that he wrote the King that in agriculture lay the future of the island.

When Diego de Velásquez invaded Cuba, he took eight horses and mares, as far as can be determined the first of a long string to be moved from the original home colony of Hispaniola. Velásquez always decked out his horse with bells to frighten savages and must have presented a gay figure to the astonished natives.

Antonio de Herrera, whose general history of Spanish occupation (1601) is one of the classics of Western Hemisphere history, tells in volume VIII the following story about the occupation of Cuba:

When the word was received in Jamaica that Captain Velásquez was in Cuba, many soldiers who were serving under Juan de Esquibel desired to go to Cuba and serve under Velásquez. Panfilo de Narváez, a well-born gentleman *("hidalgo de los cuatro costados")* headed a group of thirty archers on the trip. Narváez was a tall, graceful man with light hair and blue eyes. His complexion was fair, though becoming ruddy; he was honorable, a quick and ready conversationalist, well mannered, although in certain matters not overly discreet. Velásquez received him warmly, placing him second in command. Having the Indians well in hand, the captain sent Narváez with thirty men to the province of Bayamo, about forty or fifty leagues [160–200 miles] from Baracoa.

On this trip all the soldiers were on foot except their leader, Narváez, who rode a frisky mare. The natives along the way came out to greet them, offering them provisions because they had no gold. They were amazed by the horse, since they had never seen one before. The tricky little mare, as though she sensed the admiring onlookers, became even more mettlesome, prancing about in great style.

The Spaniards took up their quarters in an Indian village. When the natives saw how few Spaniards there really were, they decided to rid themselves of the intruders. Narváez, although not as cautious as he probably should have been, nevertheless stationed a guard and kept his horse by his side. That night seven thousand Indians gathered from all parts of the province and assembled for an attack, naked as the day they were born. They were not accustomed to fighting at night, but on this occasion fell on Narváez and his men just at midnight. After dividing into two parts, they approached and luckily found the guards asleep. Now at this point, despite a previous agreement to enter the town together, one section entered before the other. They wanted to be first to get the Christians' clothes, for they had admired them from the first and had deemed the cloth of greatest importance. With this uppermost in their minds they ran shouting for the houses without waiting for the other party.

Narváez awakened in consternation, having been caught asleep as had all the others. The eager savages rushed into the huts and being so intent on securing the clothes, neglected to attack the Spaniards. The Christians were so frightened by the tumult and commotion that they did not know whether they were alive or dead. Narváez had brought a few natives from the province of Jamaica, and they lit firebrands to see what was going on. As soon as it was light, one of the Indians threw a rock hitting Narváez in the pit of the stomach. He immediately dropped to the floor thoroughly awake. Narváez told a Franciscan friar who was in the hut with him that he had just been killed. However, with the Franciscan's assurance that such was not the case, Narváez picked himself up and between the two of them they hurriedly saddled the mare. Entirely unclothed except for his shirt, and with a parcel of hawk-bells tied to his mare's tail, he mounted and galloped around town. This so terrified the natives that they ran into the woods and neither man, woman, nor child stopped until they reached a province of Camaguey, fifty leagues [200 miles] distant. Not a Spaniard was hurt in the entire affair.

Velásquez, hearing what had happened, marched over the entire province and found it entirely depopulated with the exception of some very old and sick people.

Cuba was settled rapidly after the Velásquez expedition, and the principal occupation of the settlers became ranching and mining. Soon information of new lands encouraged the organization of other expeditions. The most successful group was led by a Cuban stockman and rancher chosen because of his wealth and ability. He put all the money he had earned on his ranch into the venture, and when the suspicious Velásquez started to relieve him of his command, he sailed a rebel. The man was Hernán Cortés. After the conquest, Cortés had an opportunity to show his knowledge of ranching in his administration of Mexico, which he handled very well. Hubert Howe Bancroft says that Cortés was not only a *conquistador* extraordinary but also a cattle king and horse raiser *par excellence*. In the beautiful valley of Oaxaca he established a cattle ranch at Matlaltzinco and a stud farm at Tlaltizapán. Mexico soon had several provinces boasting of superior horses.

When horses began going to the mainland in large numbers, the people of the islands became worried because of the difficulty in obtaining any new stock from Spain. Santo Domingo was persuaded that unless restrictions were placed on the export of livestock to other possessions, the island would suffer irreparably. A proclamation was accordingly issued prohibiting any stock that might be used for breeding purposes from being shipped. Thereafter so many complaints were received by the King from the mainland colonists, who needed foundation animals, that an order soon arrived from Spain practically annulling the prohibition.

It was a quarter of a century after the discovery before any horses were taken from the islands to Mexico. During that quarter of a century the Spaniards had firmly established themselves on the islands. Puerto Rico was stocked before Cuba, for Vicente Pinzón was granted this land shortly after the discovery. Martín de Salazar bought out his right and landed the first horses and cattle. In turn, Jamaica, Trinidad, and all the lesser islands began to graze their share, although Santo Domingo [Hispaniola] and Jamaica remained the outstanding stock centers. Many men of whom we think today as great conquerors were also capable and wealthy stockmen. Cortés himself was bow-legged from the many hours he spent in the saddle.

Admiral Pedro Menéndez de Avilés in settling Florida carried with his colony 100 horses and mares, 200 sheep, 400 lambs, and 400 pigs. He established ranches and towns almost simultaneously. Menéndez, with his usual forethought, when organizing his expedition at Cádiz, selected 117 stockmen and farmers among his colonists. Many of the necessary horses he shipped direct from Cádiz, but as it was impossible to obtain sufficient numbers there, the King furnished the remainder from the royal ranches of Hispaniola; and these horses were taken to Florida in 1565. By 1650, this district in the southeastern part of the present United States had seventy-two missions, eight large towns, and two royal haciendas extending north into present-day Georgia, from which horses spread to the Indians and the English.

The first staple exports from the New World were hides and tallow. The miners paid gold for meat and horses, and shipping men, eager to have a load on their return trip to Spain, paid gold for hides and tallow. The explorers were willing to pay exorbitant prices for horses, since have them they must, and in this way a ready market was created in the colonies which far exceeded any dream of producers in the Old World. Fortunes were made. No one has ever adequately estimated the tremendous business carried out in this fashion, but its value must have run into six or seven figures annually. Horse and cow hides became staple products around which the economic life of the Spaniards was built.

The ranchers on the islands built up an aristocracy that was envied by all Europe. With unlimited wealth at their disposal, they bought the finest horses and cattle of Spain and shipped them to their island ranches. They developed the highest type of saddle animal and originated a magnificent forerunner of the Mexican saddle, covered with silver, inlaid with gold, and studded with jewels. Because they began to purchase the best Spanish horses money could buy and ship them back to the New World, Charles suddenly realized that the laws restricting the export of livestock were not sufficient; and on March 30, 1520, shortly after Cortés landed his first horses in Mexico, the Emperor placed an emphatic embargo on the export of horses from Spain. From that date until Mendoza became viceroy of Mexico, all horses sent to the mainland came from the island ranches. They, in turn, felt the drain, and in 1525 passed ordinances designed to halt some of the outflow. Their methods were not successful, however, and the original *conquistadores* obtained almost all of their horses from the islands.

☙ *Excellent Blacksmiths*

Although the islands were the first centers of stock raising, this thriving business soon spread to the mainland, in three streams: Hispaniola to the northern coast of South America, Jamaica to Central America, and Cuba to North America. Antonio de Herrera provides valuable information concerning the outstanding breeding places in *Nueva España* or Mexico. When speaking of Mexico, he says that swine, sheep, and goats were numerous, being bred with less trouble than in Spain. This was true also of horses and cows, which were present in great numbers. Most of the Spaniards in the country, according to Herrera, lived by breeding cattle and farming, and there were droves of work horses, oxcarts, mule packtrains, livestock, and saddle horses all over the country. Matías de la Mota Padilla, speaking of the same kingdom a few years later, says that 50,000 head of cattle, more than 200,000 sheep, 4,000 mules, and an equal number of horses were driven to and sold in Mexico each year.

The inhabitants of Chiapas in Central America were singular because of their outstanding ability and inclination in breeding, training, and riding horses. In the land about Chiapas, almost from the time of the first settlement, were great droves of cattle. In certain districts, it is interesting to note, there was a worm which when trod upon by the horses entered the hoof and often caused the loss of the foot. Nevertheless, it seemed to be the general opinion that horses bred in Chiapas were exceptionally fine and were the equal of the European-bred animals.

The province of Nueva Valladolid, in present-day Honduras, was one of the greatest all-round stock countries on the mainland during the sixteenth century. In the country valleys, fresh air and gentle weather combined to create ideal climate for breeding countless herds. In the city of Gracias a Dios one of the greatest sources of wealth was raising mules, which had been originally imported from Jamaica. They were used to carry grain to Salvador, and many were sent to Nicaragua and farther south for the Isthmus trade. The ranches in Vallodolid had excellent horses, says Herrera, because the land is rocky. Being a horseman, he did not think it necessary to explain that a horse raised in a rocky country walks alertly, picking up and setting down its feet nicely, and, besides developing a tough hoof, can run rapidly over any type of country without a tendency to stumble.

Perhaps the greatest horse center on the mainland was in Nicaragua, where the natives readily learned the Spanish tongue, customs, and trade. They became excellent blacksmiths and ropemakers. They bred many different types of horses, and their mules brought them immense wealth, as they had practically a monopoly on furnishing mules for the Isthmus trade. That the trade crossing the Isthmus between Nombre de Dios and Panama was no little item can be seen when it is realized that often more than 1,500 mules would leave Nombre de Dios in one day, destined for Panama. When the great fairs were held, they would carry goods of leather, meat, rigging, grain, and other products of the country to be sold at Panama, Nombre de Dios, or other specified points.

Along the coast of Nicaragua there were many meadows, called in those days *Jahanas,* all well stocked with cattle. There were some pools of sulphur around the town of Nextipaca, or at least the water apparently came from veins of sulphur, as Herrera says they "stank very much," but the pastures around the water were luxuriant and green the year round and made horses thrive so well that no matter how poor they were when put on this feed, they recovered shortly.

By 1550 outstanding horse breeders were found at Santo Domingo, Jamaica, Chiapas, Nueva Valladolid, and Nicaragua—horse breeders whose clientele was to cover two continents and whose products are still in use.

◆ Specialized Strains

Jousts, parades, war, and races were continued in the New World much as they were practiced in Spain. Apparently in the New World, as in the Old, separate strains of horses were maintained for each pursuit. All of these activities were and have been extremely popular throughout Spanish America since the earliest times. References to the specialized types of horses required are encountered by the end of the sixteenth century. The Inca Garcilaso speaks of them in Peru. He says that the horses which multiplied so rapidly in Peru were of good breeding, equal to the best in Spain, not only the types for racing but also those for parade, for labor, and for journeys. This seems to imply clearly that at least four types were to be found. It may well be taken to indicate that there were in Peru at that time, as there were in Spain, different breeds of horses, some for racing, some for games, some for parades, some for work, and still others for war. These specialized strains in the New World undoubtedly derived from Spanish progenitors of the same class or type. Those few horses brought over now and then must have been selected breeding stock to be used by the owners in the new land, each importer selecting and bringing the type or types he desired.

It was by importing a few superior animals and crossing them on the horses already in the New World that the excellent animals spoken of by the chroniclers were bred. It is not hard to see that, because of the expense and difficulty of bringing horses across the ocean to the new land, only unusually good and carefully selected breeding stock was worth importing once horses became available in the Western Hemisphere. Care and selection were to provide a foundation not only for the horse populations of America of those centuries but also for the modern Western horse of North and South America today.

Horse Lore of the Conquest

"Although the people of this country resisted, they were soon defeated by the cavalry, which they held in great fear." Statement of an unknown conqueror quoted by José de Acosta (1590).

4. Cortés, Doña Marina, and Morzillo

☙§ *Not Only Alligators Ate Horses*

It was not long after the discovery of the West Indies until the Spaniards were spreading out in every direction. Many of the expeditions, originally organized primarily to search for gold, found instead rich agricultural lands which they colonized. Their descendants are still living in those countries.

The dapper and intrepid Alonso de Ojeda, after dancing a jig on a plank for Queen Isabella, returned to the islands for more adventure. On November 10, 1509, he sailed with four vessels, three hundred men, and at least twelve mares. While he was founding San Sebastián, the second Spanish town on the north coast of South America, he lost a mare. A great alligator came out of the river and laid hold of the mare's leg, and, dragging her into the water, ate her. Nor did only the alligators eat horses during this period. Pedro Anzurez' experience in Peru (1536) was not exceptional. When his men ran short of provisions, they drank the blood of their horses, securing both food and water by this expedient. As soon as the horses would become too weak, and die, they sold them for food. A quarter would bring 300 pesos and smaller pieces, 200 pesos. Before this expedition ended, they had eaten 220 horses that had cost almost 125,000 pesos.

Ojeda left a small colony at San Sebastián to guard his fort until his return; however, Ojeda died before he had the opportunity to go back. Those left at the town waited the fifty days which he had suggested, and when this time had elapsed, they decided to leave the sufferings and hardships and return to Cuba in the boat. A serious problem then arose: there were fifty colonists left, more than the brigantine could possibly carry. They finally decided to wait a few more days until sickness and natives had reduced their number to a more convenient size. This they did, and in a short time all who were left were able to sail on the boat. Four mares had been kept alive, as they were all that kept the Indians at a safe distance. Before sailing, the colonists butchered the horses and salted them down for provisions on their journey.

Lope de Olano had even a worse experience. While building a ship on the river at Belen, his supplies ran out. The expedition had no food and dared not kill all of the horses. The men soon reached the point that as soon as a mare would foal, they would run frantically to devour the colt.

☙ *The First Seventeen*

Although all the chroniclers tell about the horses of the conquest, there are some whose accounts are outstanding. Foremost among these primary authorities is, naturally, Hernán Cortés, for he was the first to land on the North American continent with cavalry. Cortés was born in Estremadura and was a *hidalgo de los cuatro costados.* Although trained for the law, he soon found a calling more suited to his abilities. His letters are generally terse and emphatic, as a military man's should be, but their saving grace is his sense of natural beauty. Cortés was rather well characterized by his skeptical admirer, Bernal Díaz del Castillo, who said that although he spoke Latin, he was somewhat of a poet.

Díaz del Castillo was in many ways the greatest authority of the conquest. Although he wrote in his old age while governor of Guatemala, his retentive mind and facile pen sketched all the important men as well as many of the horses with all their colors and individual qualities. Cunninghame Graham characterizes him thus: "Díaz was a brave, truthful and unlettered soldier, with all the prejudices of his race, which luckily came to light on every page he wrote, for history written impartially is apt to be as dry as pemmican, and as indigestible." His history owes much to the work of Francisco López de Gómara, although Díaz disliked everything Gómara had to say of the conquest and today his account is generally considered unreliable. Gómara told his story secondhand, as he was delegated by Cortés to write the official account as given to him by Cortés. Díaz resented it, for he felt the Gómara gave too much credit to Cortés.

The last great authority was the Inca Garcilaso de la Vega, the product of a *conquistador* of noble birth and an Inca princess. This romantic union produced a historian who had certain advantages found in no other chronicler. Not only did he see the Spaniards' side, but also he never forgot the point of view of the natives. He left Peru when twenty years of age and never returned to his native land. When he finally settled in Córdoba after military service, his retentive memory and his many chats with retired *conquistadores* gave him a unique opportunity to write the history of the occupation of the New World.

References to horses abound in José de Acosta's history (1590), and there is a passage from an unnamed writer in his work which portrays the native regard for Spanish horses. Speaking of a certain advance at the time of the Mexican conquest, he says: "Although the people of this country resisted, they were soon defeated by the cavalry, which they held in great fear. One evening when the troops were quartered near a village, several of the horses broke loose and ran, neighing and jumping through the village. The natives knowing only too well how ferocious the dogs were, thought how much more so these larger animals must be." The next day the horses were found, having taken possession of an Indian hut, which they had probably entered for shelter or to eat some corn that may have been stored in the house.

The horses of the conquest deserved the best of chroniclers, and they certainly had just that in the writings of Díaz del Castillo, the Inca, and Cortés. Perhaps the Inca has preserved the best records for actual performance of the horses, even giving the distance covered on many occasions, but only Díaz del Castillo writes of them as friends and comrades. Concerning the tragic *Noche triste* and the retreat of Cortés from Mexico City, when every horse was wounded and many killed, he wrote, "It was the greatest grief to think upon the horses and the valiant men we had lost."

Díaz del Castillo may have had, with the rest of the conquerors, a lust for gold, a blind faith in religion, and a pride of race, and at times may have been cruel and arrogant, but he had more than this. He had the courage and the ability to tell the story as he knew it, and the product is a masterpiece. From the very beginning he talks about horses. After speaking of the difficulty of obtaining good mounts in Cuba, he sets out to give a complete list of the horses embarking with Cortés for Mexico, with characteristic comments on their individual abilities. He says:

> ... the horses were divided up among the ships and loaded, mangers were erected and a store of corn and hay was put on board. I will place all the names of the mares and horses down from memory.
>
> Captain Cortés had a dark chestnut stallion which died when we reached San Juan Ulúa.
>
> Pedro de Alvarado and Hernándo López de Ávila had a very good sorrel mare, turning out excellent both for tilting and for racing. When we arrived in New Spain Pedro de Alvarado took his half either by purchase or by force.
>
> Alonzo Hernández Puertocarrero had a swift grey mare which Cortés bought for him with his gold [shoulder?] knot.
>
> Juan Velásquez de León also had a sturdy grey mare which we called *"La Rabona"* [bob-tailed]. She was fast and well-broken.
>
> Christóval de Olid had a dark brown horse that was quite satisfactory.

Francisco de Montejo and Alonzo de Ávila had a parched sorrel, useless for war.

Francisco de Morla had a dark brown stallion which was fast and well reined.

Juan de Escalante had a light bay horse with three white stockings. She was not very good.

Diego de Ordás had a barren grey mare, a pacer which seldom galloped.

Gonzalo Domínguez, an excellent horseman, had a dark brown horse, good, and a grand runner.

Pedro González de Trujillo had a good chestnut horse, a beautiful color, and he ran very well.

Moron, a settler of Bayamo, had a pinto with white stockings on his forefeet and he was well reined.

Baena, a settler of Trinidad, had a dark roan horse with white patches, but he turned out worthless.

Lares, a fine horseman, had a very good bay horse which was an excellent runner.

Ortiz the musician and Bartolomé García, who had gold mines, had a dark horse called "*El arriero*" [he had probably driven a pack train] and he was one of the best horses taken in the fleet.

Juan Sedeño, a settler of Havana, had a brown mare that foaled on board ship. Sedeño was the richest soldier in the fleet, having a vessel, a mare, a negro, and many provisions.

In this manner the rugged old conqueror recalled the sixteen horses that first sailed to Mexico, giving the name, color, and characteristics of each, and leaving a record that William H. Prescott calls "minute enough for the pages of a sporting calendar." Actually seventeen horses in all arrived at Vera Cruz when the foal of Sedeño's mare is counted.

The Grease We Took from a Fat Indian

Following the route of the preceding expeditions, Cortés' troops finally reached Mexico. The presence of foreigners on the coast of

Mexico was noted by the subjects of Moctezuma, the Mexican king. Almost as soon as the Spaniards landed, they were received by representatives of this powerful king of Anáhuac.

Cortés saw here an opportunity to display before the astonished eyes of the natives the power of the Spanish forces. He ordered the artillery to fire and the cavalry to maneuver. Then he had Alvarado on his sorrel mare cavort back and forth on the sandy shores of the beach, displaying his abilities as a rider. This strange spectacle so amazed the natives that they sketched many of the scenes to carry back to Moctezuma. There were some remarkable pictures made of the men, of the horses, and even of two greyhounds taken by the soldiers. Luckily these have been retained in Rome, where they may still be seen.

A little later when in a precarious position while camped by the River of Grijalva, in the province of Tabasco, Cortés utilized his horses in an old trick. Díaz del Castillo tells the story in this fashion:

> As Cortés was in all a very clever man, he said, laughing, to us soldiers who happened to be in his company, "Do you know, gentlemen, it appears to me these Indians have a great fear of our horses. They really think they are the ones who make war upon them, and the same with the cannon. I have an idea which will further this belief. Let's take Juan Sedeño's mare, who foaled the other day in the ship, and tie her here where I am. Then, we'll take Ortiz the musician's horse, who is *muy rijoso,* and let him scent the mare. After he has scented her we will lead them apart so that the caciques [Indan chiefs] who are coming will not hear them until they arrive and are here talking to me." We did this, just as he commanded, and the stallion scented the mare in Cortés' quarters. We also loaded a cannon, as ordered, with a large ball and a goodly charge of powder.
>
> It was about noon when forty caciques arrived, in friendly manner and wearing their rich garments. Saluting Cortés and the rest of us, they covered us with perfume and asked our pardon for what they had done, saying that in the future they would be good. Cortés responded somewhat slowly, as though angry, through Aguilar, our interpreter. He

told them how again and again he had spoken for peace, how they were to blame, and ought to be put to death. However, we were but servants of the great King and Emperor Charles, who had sent us to that place, ordering us to help and favor all who would enter his service. If they were disposed as they said they were, we would take this course, but if they were not, some of the *Tepuzque,* as they called the cannon, would jump out and kill them. The *Tepuzque* were mad because of the war made on us in the past. Cortés then secretly gave a sign to fire the cannon, which was loaded, and it thundered through the hills. As it was midday and very quiet, it made a tremendous noise. When they heard it, the caciques were terrified, and since they had never heard anything like it, they believed what had been told them. Cortés then advised them, through Aquilar, to fear not, for he had given orders they were not to be harmed.

At that instant they brought the horse, which had scented the mare, and tied him near. As the mare was tethered just behind where Cortés and the caciques were talking, the stallion looked at them, and then, scenting the mare, began to paw the ground, roll his eyes and neigh, wild with excitement. The caciques, thinking he was roaring at them, were petrified with fear. When Cortés saw that the ruse had worked, he arose from his seat, went to the steed, and commanded two servants to take him away. He then informed the Indians that he had told the horse not to harm them, since they had come for peace and were friendly.

There horses were, as Cortés himself reiterated, their companions and their salvation, and between these men and their mounts there was the strongest sense of kinship. The following incident gives a glimpse of the relationship between these intrepid men and their faithful horses.

Cortés, amid the strain of war, politics, and administration, had little time to write to the King and Emperor about his horses, but on the few occasions when he did write, we can discern more of the character of this conqueror than we could perhaps find in any other existing record. When telling of the siege of Mexico in his third letter, he writes as follows:

Our people were in no danger that day, except during the time when we left the ambush. Some horses collided and a man fell from his mare. She galloped off toward the enemy, who severly wounded her with arrows. When she saw the bad treatment she was receiving, though badly hurt she came back to us. That night she died. Although we felt her death deeply, for the horses and mares were our salvation, our grief was less because she did not die in the hands of the enemy, as we had feared would be the case.

As Graham says, Cortés looked on the mare as a friend and companion, and therefore was thankful that the last words she would hear spoken would be in the tongue she had heard, and no doubt in a vague way understood, since the day she had been foaled.

Francisco López de Gómara, the Catholic father who wrote the story of the conquest for Cortés, recounts a very interesting tale concerning a vision seen by the Conqueror's men. Cortés, according to Gómara, when he heard of the vision, used it as a means to encourage his men to fight. The apparition appeared as a mounted man, in the form of San Diego, or Saint James, the patron saint of Spain. This presence drove back the fighting natives and won the battle for the Spaniards. Díaz del Castillo gently ridicules Gómara on this point, stating that although he took part in the battle, he did not see the vision. In fact, he continues, he had never heard the story until he read about it afterwards, in Gómara's book. Then, with the thinly veiled sarcasm the man of action so often has for the man of letters, especially it seems if he wears the cloth, he added that perhaps he was too great a sinner to be allowed to see the glorious apostle.

Gómara's account of the battle and the appearance of San Diego is most interesting. He speaks of the combat and of the horde of natives who crowded the Spaniards until "they were in difficulties and imminent danger, since they had no room to use their artillery or their cavalry to open a way through the enemy."

While they were hard pressed and about ready to seek flight,

Francisco Morla appeared on a dapple-grey horse and attacked the Indians so that they were thrown into disorder.

The Spaniards, thinking Cortés had come up [with the cavalry] and as there was now room, charged the enemy and several were slain. Then the horseman left and the Indians threw themselves upon the Spaniards and pressed them as closely as before.

The horseman returned immediately and joined our men. He attacked the enemy and made them retreat. Our men utilized the advantage given them by the man on horse-back and hurled themselves on the natives, killing and wounding many. As soon as the tide turned the horseman again left.

As the Indians did not for a while see the horseman who caused them to flee in terror and confusion, thinking him a centaur, they again attacked with heathen audacity, treating them [the Spaniards] worse than before. The horseman reappeared now the third time and dispersed the Indians, terrified and suffering losses. At the same time, the foot soldiers attacked, wounding and killing [the natives].

Gómara up to this point in the story implies that the horseman was Francisco Morla on a dapple-grey horse. But Morla's horse was dark brown; therefore, if it were he, he must have been riding the horse of either Hernández Portocarrero, Velásquez de León, or Diego de Ordás, as they were the only men among the cavalry whose horses were grey.

Gómara continues:

Cortés then arrived. . . . they asked if he [the man on horseback] had been one of his men. Cortés replied that it was not, as he had not been able to get there sooner. So they concluded it was San Diego the Apostle, patron saint of Spain.

Thereupon Cortés cried, "Forward, comrades, for God is with us and the glorious San Pedro [Saint Peter]."

This might seem to indicate that Cortés was confusing Saint James and Saint Peter. Perhaps, since Saint Peter was his patron, he felt he could not diplomatically commend Saint James for the assistance; so by giving the credit to God and at the same time mentioning the name of his own patron saint, he forestalled any future difficulties.

> Saying this he and his men dashed among the enemy, driving the Indians before them out of the maze of ditches to a place where the lances could be freely used. . . . The Indians . . . fleeing into the dense forests, scattered in all directions. . . .
>
> Everyone declared they had seen the rider on the dapple-grey, three different times fighting against the Indians, as has been stated above, and that it was San Diego our Patron saint.

Díaz del Castillo's account of the same battle, as might be expected, does not tally with Gómara's. Díaz says that during the fierce battle "Cortés and his horsemen did not appear, although we wished for him and were afraid some disaster had overtaken him." They fought most of the day with no help but their own arms, and toward the end of the account of the battle, he states: "Just at this time we saw our horsemen, and as the mass of Indians were wildly attacking us they did not see the horsemen approaching from the rear, and as the ground was level and the cavalry good, with horses well trained and fine gallopers, they soon reached the natives and speared them as they desired."

This is a typical Díaz del Castillo statement—"speared them as they desired." It tells so much. In another place he says, "as was convenient at the time." Another laconic sentence: "After the battle we seared the wounds of ourselves and our horses with the grease we took from a fat Indian." Díaz del Castillo was almost without a particle of humor but most human.

He continues the story of the vision:

> As soon as the horsemen had dismounted in the shade of some trees and huts, we returned thanks to God for bringing us victory. As it was the day of Our Lady, we afterwards gave to the town the name of *Santa Maria de la Victoria,* because of the great victory won on Our Lady's Day.

Díaz specifically differs with Gómara concerning the appearance of the vision of Saint James. "It may be as Gómara says, that the glorious apostles *San Diego y Señor San Pedro* came to our aid and I being only a sinner was not worthy to see them. What I saw was Francisco Morla on a brown horse who came up with Cortés."

Thus we have the two versions of the story. Possibly Gómara was a little skeptical of the tale as told to him and so added Morla at the start to appease doubters. On the other hand, either he or Cortés, who told him of the incident, may have had a failing common to storytellers, and embroidered the tale as it progressed by bringing in the saints.

Whether it was Saint James, Morla, or just a yarn is now of little import, but it does illustrate the many ways horses were utilized and likewise the adaptability of the Conqueror.

☙ *Tenochtitlán*

All along the way toward the imperial city of Tenochtitlán, which has become Mexico City, the natives gathered and discussed the strange new creatures which had so suddenly descended on their land. The Zempoaltecas said that the horses were so ferocious that the Spaniards had to put bridles on them to keep them from devouring humans. It was commonly believed that they ate the metal bits. The Indian allies of the Spaniards told how the animals could run as fast as deer, nothing being able to escape them. Whenever the horses would neigh, the natives, quaking with fear, would run for feed and water—they would feed the horses, since they were even more afraid not to care for them. The native fear of horses was the most effective weapon the Spaniards possessed, and they encouraged the Indians to believe that the horses were gods.

The inevitable happened in the battle held between the Tlaxacaltecos and the Spaniards, when the natives, probably accidentally, killed a horse. The horse was decapitated by a single stroke. Later, at Zempoala, Cortés' chestnut stallion was killed, and he took the fine dark horse belonging to Ortiz, the musician, and Bartolomé García. After a fierce battle the Tlaxacalans were defeated and thereafter gave their support to the Spaniards, sending many warriors with Cortés to fight against their traditional enemies, the Aztecs.

Riding his new black horse and heading his troops, Cortés entered the island city of Tenochtitlán before the eyes of tens of thousands of astonished native inhabitants. The Mexicans believed the Spaniards to be sons of their famous god Quetzalcoatl, and consequently, since they, too, must be gods, housed them in the palace of Atzayacatl, where Emperor Moctezuma had his principal shrines and idols. A large number of servants and priests were put at the service of the Spaniards. These servants were instructed not to forget to have a sufficient supply of fresh green fodder for the horses. The Spaniards always rode on horseback while in Mexico City, even using their horses when they went to Moctezuma's residence, which was not much farther than across the street.

It will be remembered that Cortés sailed from Cuba a rebel, against the orders of the governor. He had not been long on the mainland when Velásquez, the governor of Cuba, sent Panfilo Narváez to bring him back to Cuba. When Cortés heard that Narváez had arrived at Vera Cruz with a fleet, he decided he had best handle this emergency himself. He left Pedro de Alvarado in Tenochtitlán as commander during his absence, and with some of his best soldiers and freshest horses started for Vera Cruz.

When Cortés and his little army approached Vera Cruz, Gonzalo de Sandoval, a lieutenant and personal friend of Cortés, sent two Spaniards with dark complexions to spy within the encampment of Panfilo Narváez. These two Spaniards disguised themselves as Indians and, carrying baskets of native fruits, mingled with the enemy.

Salvatierra, one of Narváez's officers, arrogantly ordered the disguised Spaniards to bring some grass for his horse. The Spaniards, concealing their personal feelings in the matter, procured the feed. When they had returned, they remained in a squatting position while in the presence of the Spaniard, since this was the custom of the natives. Salvatierra paid his unknown countrymen for their services with a string of cheap yellow beads. How angry he must have been when he later learned how he had been tricked!

The two disguised Spaniards were allowed complete freedom in the camp, and when night came, while one was on guard, the other saddled a horse. Then they quietly stole to the patio where the saddles were kept, and after a few minutes' work hurried away at full speed. Luck was with them, as they came across another horse that was picketed and grazing along a little stream by himself. With two horses now, they hurried back to Sandoval, reporting all they had seen, heard, and done.

Cortés, upon learning that Narváez had ninety horses, had his allies, the Chinantecas, make copper spears and explained to them how to hold the spears if attacked by cavalry, butts to the ground, heads about one foot apart, some three and one-half feet from the ground. When the inevitable battle occurred, with the help of God and a little ingenuity, Cortés and his few men defeated Narváez, though the latter had the largest expedition which up to that time had been collected in the New World. Andrés de Tápia in his *Relación* tells us the story which not only explains what the two disguised men had done in Narváez' camp, but also accounts for Cortés' astounding victory over much superior forces. Sandoval's man had cut the cinches on the saddles of Narváez' troop practically in two, so that when the battle started, the cavalry soon became footmen and, in the dark, Cortés and his few soldiers came away with the victory. Cortés had gained sorely needed supplies, horses, and fresh men for his conquest.

When he had persuaded the defeated men to accompany him, Cortés returned to Tenochtitlán, where Alvarado's ill-advised actions had created a delicate situation for the Spaniards. The Mexicans let the Spaniard enter, but they soon found that they were little better than captives within the city. The many outrages of Alvarado had made it unsafe for them to remain in the city or to travel in small groups. After some deliberation, Cortés decided to leave the city quietly during the dark of night.

Plans for the exit were carefully laid. First the booty—the treasure of Axayacatl—was distributed by Cortés. Each individual soldier was given a portion. The horses received shares equal to those given a foot soldier, and Cortés, as befitted his rank, received extra, and a fifth part belonged to the crown. The treasure proved no little problem. Cortés placed six of the slowest horses and a mare heavy with foal at the service of the royal officers, to be used as pack beasts, the best horses naturally being reserved for the fray.

With the greatest of caution, at midnight on June 30, 1520, the proud army, which only a few months before had so triumphantly entered the city, resplendent in their shining armor and colorful clothes, quietly stole from their quarters. They were heading for the country and trusting in God to lead them out safely.

Most of the cavalrymen, led by Cortés and supported by Spanish and Tlaxcalan infantry, headed the procession, carrying the homemade bridge which was to furnish passage across the breaks which cut the causeways leading to and from the island city. Part of the cavalry under the doughty Sandoval was to guard the treasure and the prisoners. The remaining few horsemen, under Pedro de Alvarado, were to cover the retreat of the fugitive army. To produce as little noise as possible during the retreat, the Spaniards wrapped native cotton cloth around the feet of the horses and the wheels of the cannon.

In spite of their precautions, they were detected, and in an incredibly short time found themselves surrounded by furious Mexicans. The bridge, when placed at the first gap, slipped from the causeway into the lake, hindering them from crossing the remaining cuts in the causeway on the road to Tlacopán. From every side the Spaniards were attacked. The cavalrymen tried to escape by jumping across the cuts, but in the dark of the night and crowded by their soldiers, most who attempted jumping fell, horse and all, into the blood-stained waters. According to the account, it was not long until the bodies of the slain filled in the causeway to the level of the road; and those men left, although greatly reduced in number, succeeded in reaching the village of Tlacopán. Don Hernán, wounded and exhausted, waited in this village for those few men who might straggle in. Many of his bravest and most loyal friends were no longer at his side. Afflicted by his grief and pain, both physical and mental, and hardly knowing what he was doing, he turned back to search for those left behind. After having walked for some time, he met Pedro de Alvarado, muddy and covered with blood, still carrying a spear in his hand. Four Spanish and eight Tlaxcaltecan soldiers accompanied him. All were wounded. Alvarado was particularly sad. He had left his beautiful sorrel mare dead in the canal.

Others listed among the missing were Velásquez de León with *La Rabona,* that sturdy grey mare; Morla, with the dark brown stallion of which he was so proud, since it was the fastest horse in the army; and Lares, who next to Sandoval was the best rider in the army, with his beautiful bay horse which was so well reined. These and many more brave soldiers and horses paid dearly with their lives for the temerity of the enterprise. Botello, the astrologer, who had proposed the retreat, and his stallion were both missing. Díaz del Castillo says that Botello had predicted that he would die at the same time as his horse. If he had, he proved a good prophet.

The Plains of Otompan

The dawn came following that "sad night," the *Noche triste,* and the remains of the little army gathered to take stock. Truly their state was deplorable. All were wounded, some still bleeding. By the grace of God and Sandoval's clever leadership, the horses which carried the gold were there. Of all the steeds which Narváez and Salcedo had brought when they came after Cortés, as well as those seventeen which were first to tread on the soil of the North American continent, there remained only twenty-three, most of them wounded.

To Don Hernán Cortés, the dreams of conquest which he had cherished, the glory, the prestige, his best friends, all seemed lost. Fortune, so long his friend, had turned its back on him for the first time. Tears, despite his efforts to stop them, appeared in his eyes.

When the sun rose, the Mexicans once again commenced the onslaught. They followed hard at the heels of the Spaniards, throwing rocks, shooting arrows, and shouting, "Not a single one shall escape!" The Spaniards who were severely wounded were tied on their horses and the retreat continued. To add to the difficulties, they missed the trail which led them to the friendly land of their Tlaxcalan allies. Hunger struck next. Prickly pears and *capulines* provided the only food available. Their pursuers harried them constantly. They had killed four more Spaniards, and were waiting for the opportune moment to deliver the final blow. Weariness and discouragement filled the stout hearts of the slowly retreating and broken men. Even the horses, half-dead with exhaustion, seemed discouraged. Martín Camboa's horse died; and while at other times its body would have been decently buried, now it was eaten by the hungry Spaniards. Truly they were in desperate straits—even Cortés was too weak to refuse.

On the following day the most disheartening spectacle of the conquest presented itself to the weary, although no longer hungry, army. On the Plains of Otompan and on the small hills which surrounded the plains, across the very road they had to follow, were thousands of savage warriors waiting for the kill, resplendent in their colorful feathers and showy dress. The fierce hostility of the

Indians seemed now to be fanned into a feverish heat as they gathered to rid themselves of the hated Spaniards. The battle, if it might be so called, which followed was the most terrible of the conquest. The Spaniards, feeling already doomed, fought with reckless abandon, their only desire to meet death, the sooner the better. With bandaged hands they grasped their weapons, ignoring the old wounds which opened with the violent exercise. Despite the apparent hopelessness, once they started fighting, it seemed that strength and courage replaced their weariness. The horses, evidently sensing the mood of the Spaniards, began to prance, perk their ears, and raise their tails, and generally show that they understood the coming struggle and were ready to fight with their masters.

In the midst of the battle Cortés was wounded. A rock struck him on the head, and soon thereafter an arrow wounded him on the hand. His black horse, the proud Morzillo, was struck in the mouth at the same time by another arrow. Cortés dismounted; and Morzillo, finding himself without his master and mad with pain, furiously attacked the enemy, kicking and biting the Indians, who gave way panic stricken at this unparalleled onslaught. The two Spaniards who were sent after the charger were scarcely able to control Morzillo and bring him back. If they had not succeeded in their mission, Morzillo would not have become an immortal, as the Indians would probably have killed him. However, Morzillo was spared for a more glorious fate.

The cavalry was responsible for the ultimate victory on the Plains of Otompan. The perspicacious Cortés boldly decided to attack the Indian leader himself, though he was surrounded by countless warriors. Mounting once again the faithful Morzillo, with a handful of picked horsemen, he fought his way toward the chieftain. It was Cortés who knocked the Indian chief from his palanquin with his spear, and Juan Salamanca beheaded him. This broke the spirit of the natives, who soon retired in confusion, unknowingly leaving a practically defeated enemy. Thus ended the historic battle of Otompan, and the Spaniards were saved from what had seemed certain death. Had it not been for the horses, and Morzillo in particular, the outcome would have been very different, and if it had, the conquest might have been delayed indefinitely.

While the battle on the Plains of Otompan was in progress, Juan de Yusto y Morla, with supplies, forty-five men, and five horses, left Vera Cruz, heading for Mexico City. They went by Tlaxcala, where three hundred allies joined them. Without knowing anything about the happenings during the *Noche triste,* Morla entered the territory of Anáhuac. Taken completely by surprise, the entire group was captured by a party of Mexicans, who sacrificed them to their gods in Texcoco. The horses, fittingly enough, were sacrificed beside and with their masters, and their hides were carefully stuffed with dry grass and placed in the main temple. A similar fate had already befallen the horses killed during the *Noche triste.* The heads of these horses had been hung in the temples carefully alternated with those of the Spaniards. Even in death, horse and master were not separated. The Indians thought that if the invaders should return, their horses would be frightened and run away upon seeing the heads of their dead kindred, and the thrice-hated Spaniards would not be able to handle them.

Cortés Has His Day

When Cortés, wounded, defeated, and discouraged, finally reached Tlaxcala, he had only twenty horses left. Now once again fortune was to smile on him. Unexpectedly he received a supply ship from Cuba. The supplies had been sent by Diego Velásquez and were intended for Narváez, whom Velásquez naturally enough presumed victorious. The ship, commanded by Pedro Barba, carried soldiers, weapons, a horse, and a mare. Barba was skillfully drawn into a net; and he, with all his supplies fell into Cortés hands. Shortly afterwards three more ships arrived; all these were the property of Francisco de Garay, conqueror of the Pánuco. The first ship, commanded by Rodrigo de Marejón, brought guns, ammunition, men, and a mare. The second, under Ramírez and called *El Viejo,* well equipped with supplies, arrived safely with fourteen horses, plenty of food, and reinforcements. Immediately after it came a third under Miguel Díaz de Auz, which brought seven additional horses and more supplies. Díaz de Auz' men were all veterans of many battles with the natives, and they had clothing padded with cotton as a protection against arrows. Consequently they were nicknamed *"los de las albardas"*—"the cotton boys." This armor protected its wearer from the powerful Indian arrows better than the second-grade metal plate worn by most of the Spaniards.

Cortés was jubilant. From a pitiful handful, he had suddenly once again acquired an army. With typical sagacity, he decided to march immediately on Mexico City. The first step was to review his troops, which he did in the main patio of the principal Indian temple of Tlaxcala. Wearing clean armor, shining like silver, under a crimson velvet cape, and mounted on the coal-black Morzillo, he must have been a striking sight as he reviewed his newly acquired cavalry, which now included some forty men divided into four squads of ten each. The army was soon making the return trip to Tenochtitlán, accompanied by innumerable Tlaxcalan allies.

In Texcoco, the welcome news of the arrival of still another ship reached the Spaniards. This ship was owned by Juan de Burgos, who offered for sale gunpowder, crossbows, muskets, and three horses. Cortés sent some of his men to buy everything they could. Before they had completed their mission, even the men—including Francisco Medel, the pilot, and all the crew—had joined the Cortés expedition. Furthermore, another ship with more supplies reached San Juan de Ulúa. This was probably that of Solís de la Huerta, whom Cortés had dispatched to Jamaica for supplies. Through these fortunate arrivals, the lucky Cortés was once again well equipped and supplied.

In the final siege and capture of Mexico City, there was not a single move in which the horses were not of utmost value. One of the attacks which Don Hernán himself led against Mexico City was repelled with heavy losses. Fifty-three Spaniards, a large number of Indian allies, and five horses were captured. The allies were taken to the various temples to be sacrificed. The fifty-three whites and the five horses were taken to the main temple to be honored by a special ceremony.

The next morning Cortés, furious because of his losses, viciously attacked the enemy, leading the army in person again. Fortunately he had left Morzillo in camp to rest that morning, for the Indians succeeded in killing the horse he was riding and captured Cortés himself. They did not kill him, because they wanted to take him to one of the temples and sacrifice him there. *"Malinche, Malinche,"* the Indians shouted when they had Cortés in their hands, for that was their name for him, derived from the name of his charming Indian mistress, Doña Marina. Christóbal de Olea made a solitary, brave, but rash attack on Cortés' captors. Although Cortés did his best to help, Olea and his horse were soon killed. Another rider tried to reach Cortés to give him a horse on which to escape, but he, too, fell with an arrow in his throat. Finally, some Tlaxcaltecos succeeded in freeing Don Hernán, who escaped on the horse of the dead soldier.

Seeing his forces weakened day by day because of the disloyalty of his allies, the Mexican leader Cuauhtehmoc, Moctezuma's successor, sent messengers to the near-by towns, announcing that the gods promised an early victory. To convince the Indians that the invaders were mortals and his statements correct, the messengers carried with them the heads of two horses. He hoped in this way to inspire them to a final victory.

In one of the last attacks on the city, a rider going at full speed threw his spear at an Indian, piercing him from one side to the other. While the Spaniard was trying to extract his weapon, his horse stumbled on the body and fell. Immediately a shower of missiles rained over the soldier, who was instantly killed. The Indians captured the wounded horse and sacrificed it.

It was not strange to see the Mexicans with horse tails tied to the headdresses they wore in battle. This was to show their bravery in attacking and killing the terrible beasts of the Spaniards. Most of the horses killed were slain with knives and swords taken from the dead Spaniards and tied to long sticks. Many were also killed with Spanish spears which the Indians had captured in battle.

When one reads the story of the conquest, he cannot but be impressed by the important place the cavalry occupied in every move. It was always at the head and the rear of the expeditions to protect the infantry. When they were not covered with mud and blood (which was seldom), the Spanish horsemen in their armor reflected every ray of the sun, and presented to the astonished eyes of the natives (in either case) a fabulous and fearful apparition. The Tabasqueños called the horses "tequanes," meaning "monsters." The Tlaxcaltecos thought the horses were deer, which by some magic power permitted themselves to be mounted. They believed that these animals flew and even talked—even Moctezuma had the same belief when he first saw the horses. The Spaniards themselves were scarcely less superstitious, for one time they declared that the Indians' witchcraft had made five horses fall sick. If one factor could be singled out which contributed most to the successful conquest of Anáhuac, surely it was the horses.

◈§ *Hippomorphous Deity*

The difficulties which Cortés had to face on his trip from Mexico to Honduras defy exaggeration. No one knew better than he that the success of his desperate intentions lay ("después de Diós") upon his horses. Consequently, in his fifth letter references to horses abound. He never failed to note in his letters the conditions and performances of his horses and anything unusual that might occur to them—this from a general who was preoccupied with all the worries and cares of a conquest. But then he knew that he was writing to the "Prince of Horsemen" and one who had always admired a good horse. And it is from this fifth letter that we get our first hint of the singular story of the horse which became a god.

Although Bucephalus, Sharatz, Babieca, Rosinante, Steeldust, Eclipse, and Mancha y Gato were all heroes of their race, another figure should be added to this immortal roll: Morzillo, for he had the strangest destiny of his kind.

When Cortés found it necessary to go to Honduras in 1524 to reason with a certain rebellious lieutenant named Olid, he rode his ever faithful charger, Morzillo, who had performed so nobly during the siege of Mexico. By his side rode the intriguing Doña Marina, on her last adventure with Cortés, whose wife was soon to come to the New World. The colorful procession leaving Mexico City consisted of a train of courtiers in brilliant clothes, and musketeers and crossbowmen clad in buff jackets and half-armor, with their steel helmets catching the rays of the sun. In the rear swarmed the ever present herd of swine, for the Spaniards always carried their meat on the hoof.

Picturesque as this motley crowd must have been, the journey was probably the most arduous military expedition of its kind man has known. The only guide was a crude compass, supplemented by an occasional Indian captured by the way to lead the party to a village and food. Since they had slight idea where they were going and less what awaited them, they relied almost wholly on their horses. The story of the terrible trudge through hundreds of miles of tropical jungle fills one with admiration for the enormous courage and the fanatical faith they had in their ultimate success.

One day during the march Cortés came upon the lovely valley of Tayasal, whose green slopes led gently down to an island-studded lake. Upon its quiet water the tall white walls of a reflected island city glimmered in the afternoon sun, doubtless bringing to the veterans memories of their first glimpse of Tenochtitlán nestled in the tarn at Mexico. Around the shores of this sequestered lake countless deer grazed, still happily unacquainted with the Christian idea of sport. Cortés and his men, with typical enthusiasm, gave merry chase, disregarding the afternoon sun, shooting and lancing the deer until satiated with the sport. The chase was not without its ill effects. The horse of Palacio Rubias died; Díaz del Castillo tells us that it was because the fat in his body melted; but since it was early afternoon in the tropics, sunstroke was probably the cause.

In the island city of Tayasal lived the Maya tribe of Petén Itzas, whose history included such fascinating details as the abduction of a bride, subsequent flight from northern Yucatán to the interior of Guatemala, and final settlement in the impregnable city of Tayasal. Chief among its deities was Hubo, whose main characteristic was a yawning aperture sufficient to receive the blessings of a human sacrifice. The chosen victims obligingly implored the favor of the gods while being roasted alive, their friends dancing around the god, mingling their chants with the victims' no doubt frantic supplications. The gods of the Itzas were supposed to join the dance. If they did, their intimacy with mortals seems ill advised, as apparently this familiarity bred a certain cynicism in the people; for if an oracle were not fulfilled, the priests could ostracize the god. Such treatment of a deity may seem sacriligious, but it is never wise in matters of this kind to criticize customs hallowed by time. Certain animals also were objects of the Itzas' veneration, among which were the deer.

Thus, the island people were living quietly, tending to their sacred deer and honoring their gods, when strange rumors began to drift in. Then they beheld with their own eyes a most bewildering sight. Awe-inspiring creatures were pursuing their deer and slaying them as they spat thunder and lightning. Although the Itzas were shocked at the inhuman barbarity of these new creatures who killed their pets, they decided that discretion was the better part of valor and invited them to visit the island.

Cortés and his company had made camp near the lake to rest after the fatigue of the chase. Canoes were seen approaching presently, and an invitation was extended by the Indians to visit Tayasal. Cortés, against the wishes of his companions who feared a trap, took twenty men and Morzillo, who was worth the whole twenty as a pacifier, and went to visit Tayasal. Here he was, as he says, well received *("donde nos recibieron bien")*. The Christians, who in turn were shocked by the inhuman barbarity of the natives, had a suitable sermon preached by the padre, pointing out the error of the Maya ways and starting the Indians on the road to salvation. This introduction to theology was gravely received, and Cortés and his padre rejoiced to see new converts. When night drew near, Cortés felt it best to leave. However, Morzillo had run a splinter in his foot and could not be taken along. Although Cortés disliked leaving his faithful animal, prudence told him that he must get back to his men before dark. Cortés tells the story simply. He says Morzillo "got a splinter in his foot and could not be taken. The chief promises to cure him but I don't know if he will succeed." As he was destined never again to set foot in the province, Cortés never knew of Morzillo's ultimate fate—or of the futility of the sermon.

The Itzas were awed by the responsibility entrusted to them when Morzillo was placed in their care. They had seen the horse flashing over the meadows, spouting incandescent death with every leap. Since obviously here was a very god among gods, they named him "Tziminchac," god of thunder and lightning. Eager to gain the favor of the new god, they decorated him with garlands of flowers and brought chickens and other delicacies for him to eat; in short, they treated their hippomorphous deity with every honor within their ken. Morzillo, unfortunately, either from grief over losing his master or perhaps from the change of diet, wasted away until only the bones of the apotheosized charger remained.

About a century later Father Bartolomé Fuensalida and Juan de Orbita made a missionary trip to the Petén Itzas, thus becoming the first Europeans to visit the Tayasal Valley after Cortés. They were well received by the natives and the day after their arrival were escorted through the city. They were not a little surprised to see temples of a size equal to any in the Christian province of Yucatán. Twelve would each hold over a thousand people. To the utter amazement of the padres, in the center of the largest temple there stood an enormous statue of a horse seated on his haunches.

No sooner did Padre Orbita catch sight of the idol (says Padre Fuensalida) than it seemed as though "the spirit of our Lord descended on him, and, carried away with a fervent and courageous zeal to the glory of God, he seized a stone, and clambering to the top of the heathen idol, battered it to pieces, scattering the fragments over the floor of the Shrine." The Indians, infuriated by such disrespect for their graven image, cried, *"Matadlos, mueran en recompense de la injuria que le han hecho"* ("Kill those who have destroyed our idol")! Fuensalida, with admirable courage and vigor, replied, "Don't you know, Itzas, that the idol which you adore is nothing but the image of an unthinking beast?"

The natives knew that the statue was not the god as well as the padres knew that their crucifix was not Christ, but this overture nearly granted the padres a coveted martyrdom. However, the chiefs restrained this impulse toward additional hospitality; and the fathers, finding further preaching of little avail, soon left the island.

Thus we have the story of El Morzillo, the horse who became a god. Even today, if you ask a native canoe-man of the city of Remedios, which has grown up on the ruins of Tayasal, he will tell you that on clear, moonless nights you may see Tziminchec deep in the waters of the lake, tolerantly receiving the worship of the Itzas while he awaits Cortés' return.

5. For God and Their King

⚜ *De Soto Remembers He Is a Gentleman*

Estremadura seemed to be the home of the conquerors of the New World. Surely it must have been left almost without inhabitants in those early days, so many *Estremaduros* flocked to the conquest, drawn by the adventurous spirits of Pizarro and Cortés, their fellow countrymen. It was in the little town of Barcarrota in the district of Cáceres that Hernando de Soto was born, an *Estremaduro* as he should have been. Yearning to follow in the illustrious footsteps of his countrymen, Cortés and Pizarro, De Soto found himself, in rapid succession, first in Panama, then in Nicaragua, and then on the coast of Castilla del Oro. It was there that he first learned the conquering game. Gonzalo Fernández de Oviedo y Valdés says (1570) that he was brought up in the "bad school of Pedrarias Dávila [his future father-in-law] in the destruction of the Indians of Castilla del Oro, graduated in the deaths of the inhabitants of Nicaragua, and canonized in Peru, according to the order of the Pizarros."

In 1532, when not more than thirty-three years of age, he led reinforcements from Nicaragua to Peru. He was no longer just a boy with his horse and sword, as he had been when he landed in the New World, but a *"famoso capitán,"* one who had fought in Nicaragua, and one of the first to participate in the conquest of Peru.

There is a picture of De Soto in the front of Antonio de Herrera y Tordesillas' history. His hair is close and wavy; he wears a pointed beard with rather long moustaches, looking, as Graham says, more poet than warrior, although he wears light armor. Herrera describes him as "rather above middle height, so graceful he looked well both on foot and on horseback, where he was exceedingly skillful, cheerful of countenance, dark in complexion, an endurer of hardships, and most valiant."

Soon after De Soto's arrival in Peru, a messenger was needed to visit the Inca ruler, Atahualpa, as there was some discussion concerning succession to the throne between Atahualpa and his brother. Pizarro chose De Soto for this trip, which had to be carried out with suitable ceremony. The Inca had previously dispatched an embassy with splendid gifts of gold and costly articles. Perhaps the Indians felt that the horses of the Spaniards could be appeased with their gold, but the result was quite the reverse; the gold increased the desire of the Spaniards, who never had enough of it. It was not long until in some cases they had to shoe the horses with gold, for this metal was easier to obtain than any other. It certainly must have provided a weak shoe that needed constant attention.

De Soto at last reached the Inca's camp, which was situated on a small plain about a league in length. The camp was on a low hill, carefully chosen, and the tents were made of fine cotton. A body of infantry came out to meet the Spaniards and to give the first recorded opportunity to see De Soto in action. The Inca Garcilaso de le Vega reports: "On seeing them [the infantry] De Soto, to let them know, if they were not his friends, that he alone was quite sufficient for them all, charged on his horse at full gallop, and stopped close to the Inca." During this, his first interview with the Peruvian ruler, De Soto must have forgotten for the moment that he was a gentleman, for he galloped up to the Inca so close that the horse snorted in the Inca's face. López de Gómara, in his record, says that "De Soto arrived making his horse curvet for bravery, or to amaze the Indians, close to the chair upon which Atahualpa sat." Atahualpa, however, did not bat an eyelash, although the horse snorted right in his face. He sat solid as rock, even though it was the first horse he had ever seen. Then De Soto, remembering that he was a gentleman, dismounted and made a deep bow. The Inca, according to Herrera, had several of his courtiers executed because they had deigned to show fear, helpfully adding that the creatures were as common in the stranger's country as sheep (llamas) were in theirs.

The Inca Garcilaso, half Peruvian himself, and anxious to make the best of the behavior on both sides, treats the whole tale with scorn. He declares that the Inca Atahualpa could not have been so ill advised as to kill any of his own men before ambassadors, nor could De Soto have been so discourteous. Reviewing this breach of diplomacy, with the ease that is characteristic of mere onlookers, or those who write at least four hundred years afterwards, it cannot be denied that Garcilaso may be justified upon the grounds of reasonableness. It is much more pleasant, though, to imagine the young Spanish captain riding into history *a la jineta,* wheeling his horse gracefully a few feet off, bringing its head into its chest by tightening the reins a bit, causing it to curvet and prance nervously before the Inca's chair, just as a cowboy in the rodeo parade. One thing, however, is certain; the charge once over, De Soto did remember that he was a representative of Spain, and, dismounting and making a deep bow, told the Inca why he had come.

Even though this episode may not show De Soto in an altogether favorable light, he more than redeemed himself when he was the only man of position who protested against the murder of the Inca Atahualpa by the Pizarro brothers.

De Soto left Peru not long after the sack of Cuzco in 1533. On his return to Spain he married the daughter of Pedrarias Dávila, who, according to contemporary accounts, was extremely beautiful and should have been, since she was marrying such a wealthy and famous man. Even so, De Soto was not satisfied to stay at home. The New World was in his blood. Nor did he have difficulty in gathering together an expedition. One hundred thousand ducats, a famous name at thirty-seven, and an adventurous spirit even in those days were sufficient to attract many men. He undoubtedly used his gold to good advantage, for in 1537 Charles V gave him permission to conquer Florida.

De Soto left on April 6, 1538, for Cuba, with his favorite horse, Azeitunero, and his bride. Which ship carried them is not recorded, but De Soto must have known the proverb of "the master's eye," especially at sea, and have taken them with him in the *San Cristóbal.* On April 21, 1538, Easter Sunday, they dropped anchor at Gomera, in the Canaries. The Governor was overjoyed to see the party and at the prospect of a ball in De Soto's honor. Everyone had a good time during the stay with the possible exception of a couple of soldiers, who, having indulged too freely in the governor's hospitality, fell overboard locked in each other's arms. Having armor on, the two quarreling men sank to the bottom of the bay like stones.

After three days' feasting, the party prepared to sail. Just before they left for Cuba, the Governor appeared, leading a beautiful damsel by the hand. Her name was Señorita Leonora de Bobadilla, and she proved to be his illegitimate daughter. Richard Hakluyt in his translation of *The Gentleman of Elvus* calls her in good old nineteenth-century English "the bastard daughter of the Earl." In any case (according to the Inca Garcilaso), she was extremely beautiful *("era de extramada hermosa").* The Governor entrusted her to Doña de Soto to be made a great lady, and with this fair maiden the Inca says De Soto left the island most content, on April 24.

De Soto arrived in Cuba toward the end of May, 1538. After the customary feasting, one of his first acts was to buy more horses. During the holidays he had plenty of opportunity to see how good the horses were, how fiery and suitable for war, bull-fighting, running at the ring, and cane-play *(juegos de cañas).* Horses were being bred in the islands by this time primarily to sell to the *conquistadores* going to war, in both Peru and Mexico. They were popular because they endured the climate better than those bred in Spain. Don Basco Figueroa was a welcome addition to the party here, bringing as he did a host of Indians and Negroes, along with thirty-six horses for himself and fifty for De Soto.

On May 12, 1539, the fleet set sail, carrying about three hundred horses, the best troops, and the finest officers yet to set out on a conquest in the New World. Nineteen days after leaving port, they dropped anchor in a deep bay off the coast of Florida, a trip that today by air takes about as many minutes. It must have been quite a sight to any natives on shore watching the Spaniards arrive. They would have seen the strangers lead the horses one by one to the edge of the boat and make them jump into the water. Then a man would jump in and crawl, dripping, on one of the fierce monsters and ride him safely to the beach. The horses, if they were not too stiff and lean after their long confinement, certainly must have torn up the sand, jumping, frisking, and generally loosening their stiff muscles.

De Soto marched northwest into what is now Georgia, a region where winters are severe and where, for some three hundred years, no gold was discovered. De Soto rested for a time at Osachile, where a favorite diversion was riding the horses up the steps of the native temple. As the troops were young and horses plentiful, this proved an interesting sport.

☙ *The Best Horse in the Army*

Two hundred and fifty years—*poco más o menos*—before Paul Revere cantered across the New England countryside, Gonzalo Silvestre and Juan López made a memorable dash for help through the Everglades of Florida. They were riding for God and their king, as well as to succor De Soto, who found himself in an awkward situation. He had wandered off into a swamp and become separated from the main body of his troops. With hostile Indians on all sides and with insufficient provisions, he deemed it wise to send for reinforcements.

De Soto knew of a certain young man who was always in the front line of battle, and though he had barely reached his majority, was *muy jinete*. This was Gonzalo Silvestre. Therefore, De Soto called Gonzalo before him and in front of his men explained what he wished done; after saying that Gonzalo might choose a companion, he gave minute directions for the perilous trip. Realizing that the mission amounted to little less than a death sentence, De Soto explained that the reason he was choosing Gonzalo was that he had the best horse in the army—*"el mejor caballo de todo nuestro exercito."* De Soto was human.

After Gonzalo had heard the orders, he turned and walked to his horse without a word. What he may have said to the animal while he was saddling is something else. Mounting Peceño, he jogged through camp in search of Juan López Cacho, one of his friends. De Soto's great name had attracted sons of the best families in Spain for his expedition to the land of the "Fountain of Youth," and many of his men were *hidalgos de los cuatro costados*. Juan was one of these, as well as De Soto's favorite page. Gonzalo found Juan asleep under a tree. Awakening him, he hold him that the chief had commanded that he should come with him on a ride to the main army for help. Juan, who was worn out with marching and fighting and who also perhaps foresaw more peril than adventure, said he was too exhausted to make the trip. Gonzalo, tired himself, replied sharply, "Soto said for me to pick a companion, and I pick you. Now either come with me or stay in God's name, for the dangers will not be the less for your absence." In this he was entirely correct. Gonzalo mounted Peceño and rode on down the path. Juan, wearily saddling, followed.

The first four or five leagues, on an Indian trail, were fairly smooth, but before they had gone much farther, their difficulties commenced. They encountered mires, streams, and lowlands, which showed that they were approaching the main swamp. Fresh Indian signs were also evident. At the edge of the swamp the trail ended; as Inca Garcilaso de la Vega tells the story, they could never have continued had it not been for God and their horses' instinct. The horses, according to the Inca, put their heads down and trailed their own scent back the way they had come. At first the youths endeavored to guide them, but then, showing a wisdom worthy of men years their senior, allowed the animals a free rein. Like bloodhounds on a warm scent, the horses moved along with noses close to the trail. When they would occasionally inhale some dust or loose grass and blow, fear scurried through the riders, who imagined Indians behind each shadow.

Gonzalo's horse was by far the best tracker, but that was not surprising to the Inca. Was he not the perfect color both for war and for peace? He was a deep chestnut, so dark that he was almost pitch colored. He had a white stocking on his near front foot and a blaze on his forehead, which he seemed to drink, as it went down to his lips *("ser bueno en estremo, porque era castaño escuro, peceño, calcado el pie ezquierdo, y lista en la frente, que bebía con ella")*.

The *conquistadores* always declared that a horse was as good as his coat, just reversing the old English proverb which says no color can be bad on a good horse.

Juan López had a buckskin with a beautiful black mane and tail. He, too, was an unusual horse, although not to be compared with the chestnut of his companion.

All through that night they kept on their way, guided by their steeds. Two days passed in this manner. The only diversion was the occasional twang of a bowstring and the whistle of an arrow. As the chronicler says, "they continued on their way with difficulties easier to imagine than to describe." They had no food other than some maize they were able to pick up at a deserted Indian village, and because they did not dare build a fire, they ate the kernels as they were, cold and hard. Three days passed while, hungry and exhausted, they rode on. Time and again they passed close to the fires of Indian camps. The savages were feasting and dancing, and though an occasional dog barked, "divine providence" closed the ears of the natives.

After having traveled in this state for ten leagues, Juan López gave out. Stopping, he said to his comrade, "Either let me sleep a moment or else run your lance through me here on the trail, for I can go no farther." Gonzalo, with a youth's disgust at weakness (he had refused the same request many times during the last few days), said, "Get off and sleep if you must, but if the sun rises and the Indians see us, all is lost." Juan, when he heard these welcome words, relaxed and fell to the ground like dead *("caer en el suelo como un muerte.")* Gonzalo dismounted wearily and retrieved his fallen comrade's lance, then stood holding the horses while he waited for Juan to awaken.

Before many moments had passed a sprinkle of rain began, followed by what seemed a cloudburst. Just as suddenly, the downpour ceased. The sun came out before the last drops fell. Gonzalo had also fallen asleep, but when he felt the sun, he awoke with a start and called guardedly to his companion. Juan did not awaken. Gonzalo finally succeeded in arousing him by beating him with the butt of his own spear. Juan, still groggy with sleep, clumsily mounted, and they started on their way.

They had barely set out when they heard the shouts of Indians. Unwittingly they had stopped on the outskirts of a native encampment, and the moment they moved they were discovered. Spurring their steeds, they galloped furiously for the swamp, the Indians close at the horses' heels, raising a terrific din.

Once in the water the horses were protected from the arrows, while the riders were guarded by their armor. Fortunately the main encampment of the Spaniards was not far distant, and hearing the clatter, a group under the leadership of Nuño Tobar came out to see what was going on. The weary messengers were soon among friends. This did not end the trip for Gonzalo, however, for he rode back to De Soto, guiding the rescue party safely through the swamp. Juan López stayed in the main camp, saying that as the general had commanded him neither to come nor to return, he would stay where he was. He probably slept.

Después de Díos

After a short march De Soto arrived at Mauvilla, an excellently fortified native town. Here he met with one of his severest defeats. De Soto himself was wounded, eighty soldiers were killed, and all the baggage was lost. The greatest tragedy "after the Christians" was that of "five and forty horses, who were no less mourned and wept for than the men, for in them was the greatest strength of the army." In the morning they butchered the dead horses, saving the skins for clothes and salting down the flesh to preserve their precious herd of hogs.

In January of 1541, early one morning De Soto was suddenly attacked by Indians. Many of his men were so startled that they ran. De Soto, ever brave, sprang out of bed, on his horse, and charged the Indians. With lance in hand, he bore hard on the right stirrup to cast his weapon. The saddle turned and he was left rolling on the ground. Being a good horseman, he had undoubtedly loosed the cinch upon retiring for the night and in his alarm and hurry had forgotten to tighten it before mounting. His peril was extreme, but a page named Tápia set him back on his horse. Superior arms gave De Soto the victory, but only at great cost, as he lost many men and horses and his camp was burned to the ground. The horses were for the most part shot or burned as they stood eating, because the owners had not found time either to mount them or to turn them loose. Garcilaso, quoting Alonso de Garmona, who accompanied the expedition, says that eighty horses were lost, of which twenty were burned. Perhaps the saddest loss in the conflict was that of Francisca Hinestrosa, the only Spanish woman, who was about to become a mother. Her companion, one Hernando Bantrita, had gone out to fight, and when he came back, he found her charred to a cinder. It was during this battle that an Indian shot an arrow that pierced entirely through a horse and stuck in the ground. This feat appeared so notable to De Soto that he had it attested formally by a notary who was attached to the army. Garcilaso, who, like many historians, likes to make a story as good as possible, adds that the horse belonged to the trumpeter, Juan Díaz, and was the largest and fattest in the army. He also claims to have known the notary in later years in Peru.

A new camp was established after this disaster. The Spaniards immediately set out making new saddles and retempering their burned weapons. They made bellows from a bearskin and used the barrel of an arquebus for a nozzle. De Soto and his men all worked in the forge shoeing horses and performing the other necessary tasks. It was indeed a plight for the most brilliant of New World expeditions, but it was brought about by a chief who would not admit defeat, one who ever pushed on in search of *el otra Perú* and a fountain of youth.

When the expedition had been away three years and had lost more than one hundred horses and over two hundred men, De Soto finally gave up and consented to start back to the coast. He kept near the Mississippi River, feeling that it would lead him quickly to his goal. Lack of salt, extreme fatigue, and continuous fighting all took their toll of men. At last he found it necessary to build a camp on the river and send to Cuba for reinforcements. De Soto himself fell ill with fever and on the seventh day, worn out with cares and hardships, died. His men, afraid that the Indians would see the body and know that the general was dead, and as there were no stones to sink the body in the river, decided to make a casket out of a tree. Choosing an oak, they hollowed out the trunk and, placing De Soto's body in it, boarded it up, making a rude ark. At midnight, they silently pushed it out into the middle of the stream, and there, in grief and with hurried prayers, they launched this ark, which floated for a minute in the swift yellow flood and then vanished from sight.

The Pizarros

Pizarro, though he tried, never could get enough horses in Peru. Since on his first trip from Panama he could take only four horses, he had to send Almagro for more. He also received horses from the new governor, De los Ríos. When Almagro returned, he brought 50 horses which he had procured and 120 of Alvarado's which he had picked up on the way back. When De Soto arrived in Peru, he also had horses, giving Pizarro something short of 300 horses all told.

If the Pizarros gave to Spain great riches, they gave to the world an even greater story of one of the most remarkable rides in the annals of history. Agustín de Zárate in his history (1590) gives the outline of this matchless pursuit. It occurred during the war between Gonzalo Pizarro and the Viceroy Blasco Núñez Vela. Although the incident took place a few years after the conquest, the horses participating were bred in the islands, very few coming from Spain.

Zárate's story reads as follows:

> We have already told in the preceeding chapter how Gonzalo Pizarro followed the Viceroy from the city of San Miguel to Quito, which is 500 miles.
>
> He followed so close in pursuit that hardly a day passed that the scouts of the two forces did not see and speak to one another. For the whole road not once did either party take off their saddles from the horses although in this, if anyone, the viceroy was more careful. If they took a short rest during the night, it was always with their clothes on and holding their horses' bridles. They never erected tents, but dozed while holding their bridles, ready to mount and ride, without ever tying

up their horses, even at night. For this use (when needed) each horseman had a small bag which he filled with sand and buried in a hole, having one end of the halter fixed to the bag. He then tramped the bag firmly to hold it from being drawn up by the horse.

Almost any article can be used in place of a bag. The Gauchos, Indians, and Mexicans all employed a bone or a bunch of grass, or if nothing else was handy, they tied a knot in the halter rope. Horses thus secured were perfectly safe so long as they did not become tangled up in the rope or frightened. Another common practice was sleeping over the rope, thus having the horse act as an alarm in case it was frightened by anything during the night.

The pursuit of the Viceroy lasted some little time. When Gonzalo saw that he would have to march through a desert country, entirely destitute of provisions, he discontinued the pursuit and returned to Quito. Counting from La Plata, whence Gonzalo first set out, to Porto, where the chase was discontinued, the distance was no less than 2,800 miles. Perhaps this was the longest and most closely contested pursuit on horseback ever made.

The Spanish Horse in North America

When spurs clicked and campfires flickered in the unfenced world that was the West . . .

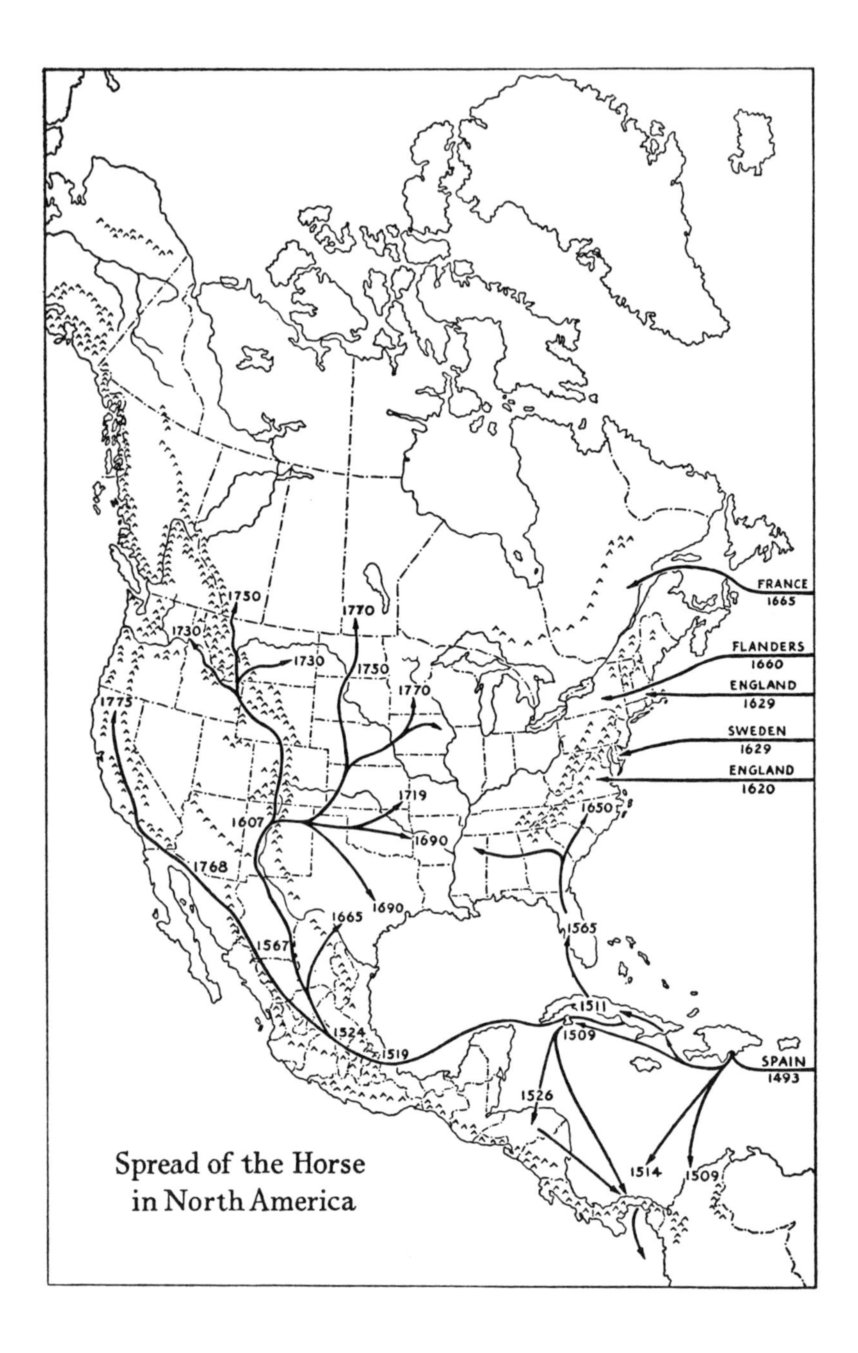

Spread of the Horse in North America

6. El Otro México

⸎ *Trail of the Buzzards*

After the conquest, Mexico began to import livestock from the island ranches of the West Indies, and it was not long until ranches appeared in New Spain, as Mexico was then called. The first "boom," however, did not occur until the silver mines were opened north of Mexico City, in the middle of the sixteenth century. Then, with the mines providing the market and the high plains an ideal pasture, the cattle business began to arise as an important industry. The horses of the conquerors never had a chance to forget the work of cow horses, for the Spaniard, even on his expeditions of conquest, almost without exception drove livestock for provisions.

The opening of silver mines about Zacatecas did not halt the Spanish advance north. In 1562 Francisco de Ibarra was appointed governor and captain-general of a new province, called Nueva Vizcaya. It was located north of Nueva Galicia and was a region where the Indians were still unconquered. Ibarra, blessed by nature with a strong body, a goodly share of intellect, and a fabulously rich uncle, became within a few years one of the ablest second-generation colonizers of New Spain. It was Ibarra who was responsible for the spread of livestock into the regions now called Durango and Sinaloa, and even into parts of Chihuahua and Sonora. The fertile river bottoms, grassy plains, and broad meadows of this country invited increase. The natives did not object, for they stole large numbers of the stock. On one foray they succeeded in getting away with 250 horses. Another time they stole Ibarra's personal mount during an attack on Guatimape.

Two years after the founding of Nueva Vizcaya, Ibarra made the arduous westward trip over the sierra to conquer and settle Topia, which he hoped would be *el otra México.* He founded the town of Cinero on the Río Fuerte, but the settlers had to retire afterward when the town was destroyed in a native uprising.

Two decades later Hernán de de Trigo determined to re-establish the lost province of Cinero. With Pedro de Montoya, who was an experienced campaigner, as captain, plus thirty soldiers equipped with all the necessary arms, horses, and provisions, he entered the province. Surprisingly enough, he and his men were hospitably received by the erstwhile rebellious natives. To their amazement they found great numbers of livestock peacefully grazing along the river meadows. Horses and cattle of all ages and of every description were in evidence. The animals left when the previous settlement was hurriedly abandoned had increased until they numbered more than ten thousand head. A better illustration to help explain the origin of the semiferal mustangs and wild cattle could hardly be cited.

Some two hundred years later, during the first half of the eighteenth century, at the time when the English colonies on the mainland of America were just getting a good start, this same country of Galicia and Vizcaya was producing large numbers of livestock annually. Mota Padilla claims that 500,000 head of cattle, 200,000 pigs, and 1,000,000 sheep were driven to New Spain (central Mexico) from these two provinces. Truly the area proved to be a stockman's paradise.

During Ibarra's trip to Topia in 1564, many horses were lost, not from the rocky and mountainous terrain, but from the cold. While crossing the summit, thirty-eight horses froze to death at night, the Spaniards awaking in the morning to find the horses dead. Ibarra's mount was found, according to Obregón, frozen erect and leaning against a tree.

Antonio Sotelo de Betanzos, camp master for Ibarra, had a most interesting experience while hunting to supply food for the new town of San Juan de Sinaloa. One day he wounded a deer which succeeded in getting away from him. The Spaniards had learned from the Indians how to find lost game—to go back the next day to the spot where they had encountered the animal and locate their kill by the trail of the buzzards who were attracted by the meat. Sotelo, having found his deer in this manner, put the haunches on the back of his saddle. This was no small job, as the horse, smelling blood, became nervous and objected. He then pursued his course homeward eight or ten miles when the way became very rough, strewn with rocks and undergrowth. Obregón, the principal chronicler for Ibarra, says: "A bear attacked him [Sotelo] furiously and boldly creating a great noise in the rocks and thickets, and making Sotelo's hair stand on end and causing him to shiver and his horse to snort in fright. To defend himself, he at once drew his sword, greatly frightened and upset at finding himself attacked by an animal so furious." To say he was greatly upset is probably putting it mildly. The animal was so near Sotelo could easily have shot it, but in his confusion he completely forgot he had a loaded arquebus.

The bear, sensing that someone was in his neighborhood, turned around and rose on his hind legs. A bear's vision, poor in any circumstances, is better when he is standing erect. Hearing a noise, he will rise and turn his head in an effort to determine just what has disturbed him.

Thus the bear Sotelo had encountered was startled when he heard the horse coming through the thicket and after a short run stood up to see what was approaching. Sotelo, seeing him standing up, arms outstretched, thought he was waiting to fight. Obregón continues: "The *maseo de campo* wanted to attack, but could not make his horse move, even though he urged it on." One cannot blame the horse, who was probably as frightened as Sotelo, but, Obregón continues: "Seeing his horse was so cowardly, he went on to the fort holding the drawn sword in his hand all the way."

The bear, when he heard the horse and rider furiously departing through the underbrush, no doubt turned and ran the other way, just as glad to depart as the other two parties to the encounter. This story is faintly reminiscent of John Hood's account of the man running with all his might and the lion with all his mane.

Betanza was governor of San Juan de Sinaloa, the city that is now called San Felipe, for eleven months. His precautions and care indicate the common procedure of that day and are a good example of a typical frontier town's alertness. "They lived with much caution and fore-thought, with guards and sentries on horseback. These he [Betanza] came to inspect and question at all hours of the day and night that they might not grow careless. Likewise he customarily made trips through the district, never resting by day or sleeping by night, alert, cautious, prepared for all that might arise. There was ever a horse saddled and ready for him that he might be the first to come to the rescue when danger arose."

For Want of a Nail

Francisco de Ibarra made a long and difficult overland trip in 1567. Going north, he ascended the Yaqui River Valley, followed up the Sonora, and then crossed the mountains eastward, arriving finally at Casas Grandes in northern Chihuahua. While at Cauguaripa (in the Sonora Valley) he was attacked by the natives and owed his life to his horses, as the following story shows.

The General, finding the surrounding country alive with hostile Indians, hurriedly organized his army for the defense of the encampment. Since the army was in a relatively secure position in a native village, the soldiers tied their horses in the center of the town and took up their posts around the edge. The natives attacked with "savage and infernal fury." The sudden hullabaloo created by the wild cries of the natives and the hollow popping of the Spanish arquebuses stampeded the horses. Since they were tied only to fallen limbs which were lying on the ground and the unsubstantial native shacks, they readily broke loose, dragging the limbs and portions of the huts. The trailing articles, combined with the rest of the uproar succeeded in frightening the animals so thoroughly, Obregón says, that "they resembled loose demons, running over, knocking down, and routing the enemy." As it was night and the dark was only accentuated by the few campfires, the horses crashing through the houses and into one another caused the whole battle to seem as "fiendish, furious, and disorderly as the depths of hell. . . . In this fashion," continues the Spaniard, giving credit where credit was due, "did God and our horses give us the victory."

One of the difficulties the colonizers always faced was trying to prevent the Indians from stealing all the horses. The natives in this territory went the whites one better, as they regularly used the horses to serve a dual purpose. They not only rode them but also ate them. This created a difficulty. They had to procure new ones more often, and the Spaniards furnished the best source of fresh stock. When the natives in the Sonora Valley became obstreperous in this respect, Ibarra decided to punish them, and organized an expedition to enter the Indian country.

During this journey he had a great deal of trouble in keeping his horses shod, for the country that the Indians fled into was "in the steep and rocky parts of the craggy and rugged sierras." The horseshoe nails brought by his men were soon gone, and once a horse lost his shoe, he was useless. The hoof would wear down until he could not walk. As a result their number decreased each day, and as they diminished, the danger to the Spaniards increased, because without the horses they were relatively helpless before their enemies. Ibarra tried to make a bellows from horsehide but, unlike De Soto, failed. He then tried to go toward the South Sea (Gulf of California) to reach ground less rocky. At this point a method was devised for making nails without fire, fashioning the points on an anvil, and leaving the heads in the shape of a right angle. When this proved successful, nails were hurriedly constructed from cinch rings, sword belts, and iron stirrups.

Obregón comments: "This was lucky, for if we had not contrived to make the nails, we should have lost our horses because of the unusually rocky condition of the ground in the sierras. The heavy rains had softened the hoofs to such an extent that in a short time they lost them and then we could not make the animals move about. Horses shod with this type of nail were good for eight days. I do not doubt for a moment we should have been lost had we not kept our horses." He was not alone in this sentiment, as it has been echoed by every chronicler of the occupation of the New World.

Strange Cattle

The first important contact of the Spaniards with the American bison, or buffalo, occurred while Oñate was effecting the settlement of New Mexico. The Spaniards had heard rumors of strange cattle, and a few had seen them. Cabeza de Vaca saw them three times and ate their meat. However, general knowledge of them was scarce. Jean Rotz illustrated his maps with caricatures of the buffalo as early as 1542, but probably the first actual engraving of the bison did not appear until 1558. Then a Frenchman named André Thevet, in a book on the Spanish conquest published in The Netherlands, showed the first picture of the American buffalo. The Spaniards were eager to go out and obtain buffaloes, which they regarded as wild cattle, to supplement their own livestock.

Juan de Oñate, soon after his arrival in New Mexico, sent some men to the buffalo country to obtain a herd for use in the new settlements to supplement his cattle. Vicente Zaldivar, Oñate's lieutenant who was sent, had remarkable experiences trying to gather the animals. The horses, trained to work with the Spanish cattle, must have worn out their hearts trying to drive the thick-skulled and flighty beasts.

After several days' travel, the group saw their first buffalo, undoubtedly an old bull who lacked the vitality to hold his place in competition with the younger bulls. The horsemen played with him for a while, but no one considered keeping him. They all visualized themselves as bringing back ten thousand head of their own. Soon after that they came to a sink which still held the remains of the last rain. Here they encountered some three hundred head of buffaloes wallowing and grunting in the mud. On the following day they must have been about one hundred miles from Pecos, and not far from the Canadian River, probably in the vicinity of Clovis, New Mexico. Here they encountered about one thousand head, and as there were trees available, they began to work on an immense V-shaped fence which was to lead into the opening of a large corral. Knowing their own half-wild Andalusian cattle, they figured that this would be the easiest way to trap the buffaloes. Unfortunately, the beasts began to drift away, and the project had to be abandoned, even though preparations to corral them were almost complete. After a holiday in honor of Saint Francis, they left that location and followed in the path of the herd. It was four days before they found another suitable place near the buffaloes, and they once more began the construction of a corral and trap.

Now the Spaniards were approximately 130 miles from Pecos, south and east of Tucumcari, somewhere near the Texas border. It is even possible that they were on the Canadian River, as they used cottonwood to construct the trap. By working hard, they completed the corral, which was large enough to hold ten thousand head of cattle, in three days. They had seen so many buffaloes during its construction that they felt the job was as good as done. They did not foresee the difficulty that lay ahead of them.

Early the next morning the camp was alive with preparations

for the expected roundup. This was an old story to most of the men and the horses, for they had been brought up with cattle and on ranches. Heading out on the plains as the sun rose, they saw approximately a mile away what seemed to be about 100,000 head of the wild cattle. Making a wide circle, the men approached the herd from the far side and started drifting them toward the corral. All was going as expected and smiles wreathed the men's faces, when suddenly something—it may have been a jackrabbit or the shadow of a hawk—startled the buffaloes and they wheeled in a body and charged back directly toward the horsemen. Zaldivar's own words give us some hint of the herculean efforts made by his *vaqueros* to turn back the tide of charging buffaloes; but here they were not dealing with ordinary wild cattle. Zaldivar says, "It was impossible to stop them because they are terribly obstinate cattle and courageous beyond exaggeration."

The men pursued the animals and did their best to drive a group toward the corral, but in vain. For several days they tried all the tricks of the cattleman's trade to get a part of the herd into the trap, but all attempts were a failure. The horses ran themselves thin trying to push the strange cattle into the corral, for they knew, having worked with cattle before, what their masters wanted. Three horses were killed trying to turn bulls, and forty were badly wounded.

Finally deciding that they were wasting their time trying to drive these older cattle, the Spaniards concluded that it would be better to capture a group of calves which they could raise. Therefore, the men got down their ropes and the plains of Texas witnessed their first roping. However, this project proved useless also, not because the Spaniards could not rope but because the calves soon died when separated from their mothers. The men at last concluded that the only way to obtain buffaloes—if there was any way—was to capture the calves soon after birth and raise them on goats or domestic cows. Thereupon the party returned to the town of San Juan Bautista to report their failure.

◆ New Mexican Horsemen

The New Mexicans never bred their horses carefully in spite of their being so justly celebrated for their horsemanship and so devoted to equestrian exercise that they have been termed a race of centaurs. They merely converted their best and most handsome horses into their *caballos de sillas,* or saddle animals.

Their horses were similar to the mustangs running wild on the prairie; although they were small, they were well formed and active. Some of the horses were marvelously trained to rein readily to either side or to stop abruptly upon the slightest touch. They would charge against a wall without shirking and even try to clamber up its sides upon command. In addition, a good riding horse in New Mexico was trained to a peculiar up-and-down gait. This gait was achieved by putting weights on a thong tied to the horse's feet. Then the horse had to lift his feet high so that the weight would clear the ground without dragging. Also, hung over his tail was a metal apron with spikes that pricked the horse just above the hocks when he walked, causing him to move his legs abruptly forward in a mincing step. The combined result of training with these devices was the desired gait.

The New Mexicans developed a riding costume almost unique, the clothes of the early California *caballeros* alone comparable to it. The outfit was always topped by a sombrero—a low-crowned hat with a wide brim—covered with oilcloth and encircled by a hat band made of tinsel cord nearly an inch in diameter. The jacket, or *chaqueta,* was marvelously embroidered, decorated with braid and barrel buttons. For trousers they wore a peculiarly shaped garment called *calzoneras,* with the outer part of legs open from hip to ankle—the border set with twinkling filigree buttons and the whole fantastically trimmed with tinsel lace and cords. A rich sash wound tightly around the waist added materially to the picturesque appearance of the *caballero.* The leggings, or *botas,* whose nearest relatives are perhaps those leggings worn by the bandits of old Italy, were made of embossed leather, embroidered with fancy silk and tinsel thread and bound to the knee with curiously tasseled garters. A *sarape saltillero,* or fancy blanket, completed the outfit. The serape was as useful as it was ornamental. In fair weather it was usually carried dangling carelessly across the pommel of the saddle, and in bad weather drawn around the shoulders in the fashion of a cape. In exceptional weather or when violent winds or riding made it necessary, the rider put his head through a slit in the middle, and by letting it hang loosely around his neck thus protected his whole body.

The horse was also decked out by the rider to match his tastes. Often the saddle trappings weighed over one hundred pounds. First the high pommel of the saddletree was crowned with silver and the back cantle heavily inlaid. The seat of the saddle was adjusted with an embroidered cushion. The saddle was covered somewhat by a *coraza* of embossed leather embroidered with fancy silk and tinsel, with ornaments of silver, and thrown loosely over the cushion and *fusta,* or saddletree, the extremities of which protruded through appropriate apertures. Then came the *cola de pato,* literally "duck's tail," which Josiah Gregg in his *Commerce of the Prairies* said could be more appropriately called "peacock's tail." This was a sort of leathern housing, (gaudily ornamented to correspond with the *coraza)* attached to the cantle, and it covered the entire haunches of the animal. The *estribos,* or stirrups, were usually made of wood, well carved, over which were fastened tapaderas. Before the tapaderas were popular, a whole slipper made from a solid block of mortised wood was used. Perhaps the most costly part of the whole outfit was the bridle, which might be entirely silver, or at least heavily ornamented with silver buckles, slides, and starts. A massive bit was used, sometimes of pure silver, but more often of iron singularly wrought. The spurs were generally iron, although silver ones were also common. The shanks of the *vaquero* spurs were from three to five inches long, with rowels sometimes six inches in diameter. One pair is in existence measuring more than ten inches in overall length with rowels five and three-quarter inches in diameter, weighing two pounds, eleven ounces. Last but not least, were the *armas de pelo,* a pair of shaggy goatskins, richly trimmed across the top with embroidered leather. These dangled from the pommel of the saddle and were used as our stovepipe chaps, to encase the legs in brush or rain. The *coraza* of the traveling saddle was equipped with handy pockets called *coginillos.*

In the eighteenth century a harness of leather was used, attached to the saddle in a way similar to the *cola de pato,* covering the hinder parts of the horse as low as the mid-thighs, with its lower borders fringed with jingling metal ornaments. However, with the beginning of the nineteenth century it was seldom found. Even without this chattering appendage, a Mexican *ranchero* of the early eighteenth century, with full outfit, made a picturesque and novel appearance.

By far the most indigenous product of New Mexico was the pasturage; the stock were never fed. They endured the most rigorous winters on the unsheltered countryside and the hottest summers in the dry pastures.

In the years following the Spanish settlement of New Mexico, gay young blades from Las Cruces to Taos would go out on the plains each year and kill buffaloes. This was a competitive sport demanding skill, horsemanship, and personal bravery. Moreover, through this practice, robes and meat were guaranteed for the winter months, and considerable youthful exuberance was given an outlet. Later there arose a class of *ciboleros,* who hunted buffaloes professionally. They presented a picturesque appearance in their leather clothing, flat straw hats, bows, and colorful arrow quivers, long lances with tassel attached, and their quick-stepping ponies. They demanded the best in horseflesh and needed it. As the pony ran alongside the buffalo, the *cibolero* would hurl the lance, the throw timed to the split second in order to hit the vital spot just behind the foreleg, which was exposed when the animal was extended while running. The moment the *cibolero* threw, the horse had to swing to one side and allow the animal to fall. A wounded animal would turn on the hunter, gore his horse, and kill the rider if he were given the slightest opportunity.

This was a sport enjoyed by the New Mexicans. It had enough danger to be zestful, enough competition to be sporting, and necessary enough that it was fully justified. The Spaniards never adopted the Anglo-American method of using the rifle—that was like shooting squirrel with a shotgun or quail on the set.

As late as the eighteen eighties, the hunters still went out to the buffalo country followed by the creaking *carretas,* as the two-wheeled oxcarts were called. The Indians driving these *carretas* could butcher almost as fast as the animals were killed. They then hung the meat on racks to make jerky, which the Spaniard spelled *charquí.* Immense quantities of it were sent to Mexico along with slaves, turquoise, *sarapes,* and wool.

◆§ *Tejas*

The story of the first livestock in Texas is equally interesting. León (Carralvo) was established near the close of the sixteenth century, and for almost one hundred years it was the closest Spanish outpost to modern Texas. In 1625 Nueva León was given to Martínez de Závala, who capably managed this frontier for two-score years. One of the best known of Závala's lieutenants was Alonzo de León, who was also one of the first men responsible for taking horses and cattle into Texas.

In 1665, one of the first recorded expeditions to cross the Río Grande was led by Azcué from Monterrey. With the establishment of missions and ranches during the next few years, it appeared that Texas was well on the way toward permanent colonization. It would be closer to the truth, though, to say that the Texas or *Tejas* Indians, as they were then called, were just waiting for the Spaniards to create a new food supply for them, because the Janamberes Indians revolted, attacked the settlements in 1675, and forced the Spaniards to withdraw hurriedly. All of the cultivated lands, stores of maize, and herds of livestock were deserted. In this raid the Indians of Texas got not only one of their first fills of beefsteak but also large numbers of horses. That they appreciated both is amply illustrated by the attack they soon launched on Río Blanco. Sixty-seven warriors crept up one afternoon and got away with all the livestock with the exception of two protected groups. One group consisted of some mares in a corral and the other a herd of mules feeding, probably clandestinely, in a protected grain field. The natives successfully drove away some two hundred broken horses.

In 1687 Alonzo de León made his first expedition into Texas. When we keep in mind the two above paragraphs, it is not surprising that he found mounted natives. On his first venture he took almost five hundred horses and descended the Río Grande to the Gulf. He made two more trips in the next two years, looking for La Salle, who was supposed to have settled on the bay of Espiritu Santo. On each trip he took between five hundred and one thousand horses and mules. In 1689, León speaks of encountering mounted Indians on his trip to Texas. In the same year he crossed the river with seven hundred animals and founded the town of San Antonio de los Llanos. During the next decade, under the guidance of Father Pedro de Villa, the district was well populated with many stock ranches. In 1690, León and Father Massenet led an expedition across *Tejas* and founded two missions among the Asinai Indians on the Neches River. Texas now became a province as far north as the Red River, with Domingo de Terán as governor.

And thereon hangs a story, of which only a few scattered fragments remain. It seems that Governor Terán died somewhere in the outposts of his province. An important personage like the governor could not be disposed of as easily as an ordinary soldier. He must be carried to consecrated ground. Summer in Texas is not the best time or Texas the best place to preserve a corpse. Finally it was decided to sew the body in a sack, and in this fashion the Governor was carried some two hundred miles on horseback in the short space of two days and laid to rest peacefully in proper state on holy ground.

The Indians would secure horses and cattle from the Spaniards, by hook or crook, and then trade them to the French and the English. When the Domingo Ramón expedition arrived in the country of the *Tejas* in the early years of the eighteenth century, they noted that the Indians had much of the white man's equipment. Upon investigation, they found that the French from Natchitoches brought articles of trade in square boats down the river and exchanged them for horses. This trade had apparently been going on since the coming of the French to the Gulf region. The French were just as eager to meet and trade with the Indians as the Spanish authorities were to prohibit any such traffic. The latter were never successful in their efforts, and the movement of horseflesh continued, even after French rule ended in North America in 1763.

7. Half Spanish–Half Wild

◂§ *Mesteños*

Around myriad campfires that once flickered at the will of a stray breeze in that unfenced land that was the West, stories of mustangs have been recounted—tales like those of the "Milk-white Steed," of "Starface," and of "Black Devil." These legends, since immortalized by that peerless raconteur, J. Frank Dobie, in his little volume called *Tales of the Mustang,* are but elusive reflections of a Horse Age, an age which saw parts of Texas prominent on the map of America only as "Wild Horse Desert."

Most of the wild horses spread in America as a result of Indian attacks on outlying missions and ranches. The natives did not care, or dare, to stop and round up all the livestock after a raid; and by the time the attack was discovered by the closest neighbors, much of the livestock had wandered away. Many times whole colonies were wiped out or habitations hurriedly abandoned, leaving the livestock to fare as they could. Mustangs and longhorns were the result. The greatest of all wild-horse ranges was in Texas, principally between Palo Duro and the Salt Fork of the Brazos, and between the Nueches River and the Río Grande.

Wild horses were every color under the sun. To the ranchmen perhaps the "coyote dun" was considered the best, but the Indian preferred the grey and the pinto, because they believed that the color would not only blend into the landscape but also take paint better.

The achievements of the Plains Indians of North America are modern and were accomplished on horseback. Pedestrians became the most superb horsemen in the world. If we feel that the automobile, train, and airplane have changed our civilization, it should be remembered that when the Indian women took packs from dogs and put them on ponies, the transition was not less for the natives. The buffalo was now an easy prey. The stolen horses, accustomed to roping and cutting the fleet Spanish cattle, would chase and overtake a buffalo more willingly than not. It took little training for roping horses to turn aside instinctively and stop at the twang of a bow string and so be out of the reach of sharp horns, as all good buffalo horses should.

The horse was both capital and a medium of exchange for the Indians, not to mention an added food supply. The only two domesticated animals the Indians kept—dogs and horses—were choice dishes. Horses were always in demand. They were obtained with less trouble from frontier settlements of Europeans than from wild bands, and if on these forays, the braves were able to take a few scalps, the expedition was regarded as just that much more successful. An Indian's wealth and family were decided by the number of his horses; next to prowess on the battlefield, skill in taking horses was regarded as the greatest virtue. As Chief Is-sa-Keep told Captain Marcy, his four sons were a great comfort in his old age; they could steal more horses than any other young men in the tribe.

It is difficult to say which Indian tribes became the most proficient horsemen. Probably the honor lay between the Comanches and the Kiowas. The artist Catlin lived among the northern tribes, and when he saw the Comanches, he said that they were unequaled in horsemanship. The facility with which they could drop to one side of a running horse, not exposing any more than the sole of a moccasin to the enemy, and shoot arrows under the horse's neck, aroused his admiration.

Like most horsemen, the Comanches were fond of horse racing, and Colonel Richard Dodge tells an interesting story about it. It seems that My-la-que-top, a Comanche chief, took "a miserable sheep of a pony, with legs like churns," and beat "a magnificent Kentucky mare, of the true Lexington blood," winning some heavy bets made with the army officers and soldiers at Fort Chadbourne, Texas. The native rider of the sheeplike pony added insult to financial injury by riding the last part of the race sitting face to the horse's tail, "making hideous grimaces and beckoning to the rider of the mare to come on." From this description it appears that the chief had a Quarter Horse in its winter coat. Certainly, if it was, it was not the last time a Texas Quarter Horse defeated a blooded horse, to the consternation of the latter's backers.

Afoot to Horseback

The Spaniards from the first did their best to keep horses from the natives. They realized that their best weapon both physically and psychologically was the horse. Once mounted, the Indian with his quiver of arrows was superior to the Spaniard with his single-shot arquebus. That the Spanish fear of mounted Indians was well founded was later beautifully illustrated on the Great Plains. The great Spanish Empire was stopped short when it reached the Plains area where the natives had horses. The European nations were not able even to protect settlements within reach of the Plains Indians, much less subdue the natives themselves. The savages held the Great Plains until repeating rifles and revolvers were introduced.

The two men who have made the best studies of the acquisition of horses by the North American Indians are Clark Wissler and Francis Haines. Wissler's work is more complete, although Haines did bring out two new facts: first, that Wissler placed horses with the Indians at a date a trifle early; and second, that the early expeditions of Coronado and De Soto were not responsible for the wild mustangs. A third point which should have been made is that the natives obtained their original horses, and always by far the greatest number, from the Spaniards or neighboring tribes and not from the wild herds. The Indians had mounts by the time the wild herds dotted the plains, and they always preferred domesticated animals to the *mesteños*. Mustangs were hard to catch and, once caught, harder to tame.

The pedestrian Indians in the New World gazed with wonder and admiration at the new people who came to their land. When they saw that horse and man were not one animal and discovered that a stolen horse could be used by an Indian, they did not rest until they possessed this wondrous creature. Actually, the Spaniards never had a chance to keep horses from the natives.

Very few horses escaped the early explorers. In expeditions such as De Soto's in southeastern North America and Coronado's in the Southwest, even though many horses were taken, they were much too valuable to lose; moreover, there was little danger that if one did get away, he would stray far, for he would not leave his companions. When one studies the records of these early trips, he finds almost no trace of any horses escaping. It is true that many were killed and some died. Those that were too weak to travel and fell behind were immediately dispatched by the Indians, who regarded them as more dangerous than the Spaniards.

Close on the trail of the exploring *conquistadores* came the padres, establishing missions and gathering the native populace into *rancherías* where they could be converted. Fathers and neophytes had to eat; therefore, gardens were planted and cattle and horses brought in. A difficulty arose at once. Who were to be the *vaqueros?* Since there were not enough Spaniards available to be cowboys, the only course was for the padres to teach the natives. Thus in the missions we have the first and primary source through which natives learned horsemanship and obtained mounts. As an example, let us look at the work of Padre Kino in Pimería Alta. His accomplishments as a stockman would alone have made him famous. His ranches were so prosperous that he became a cattle king and for several years supported the missions in Baja California with his cattle. When the mission of San Xavier del Bac was established, Kino took 1,400 head of cattle there from his home ranch at the Mission Dolores, all driven by his Indian *vaqueros.*

As the missions grew, so did the number of native *vaqueros* increase. Some would rebell from time to time and run off to their tribes with stolen horses and cattle. With these runaway Indians as teachers, the rest of the tribe required little time to become equestrian, and soon they were slipping down to the missions to steal more horses. The civil government foresaw the dangers in the revolts of Indian *vaqueros* and the stealing of the mission animals and so forbade the padres to have natives as *vaqueros.* However, the fathers had no choice, since food was essential for their *rancherías* and there were no white men available for the work.

A second source from which Indians obtained horses was the half-breed trader, called by many names but generally referred to as a *Comanchero.* These often illicit traders would gather together a few horses and go far back on the Indian frontier to trade with the natives. Many of them could speak the native tongue, and if their customers did not know how to use a horse, the traders would take time to teach them. Again and again the Spanish colonial administration passed laws forbidding trade in horses with the Indians, but the laws did not stop the commerce.

Thus the Indians learned to ride and handle horses both at the missions and through renegade traders. Rapidly the Indian became an expert horseman, and while he did not handle his horse with the brusque impetuosity of the Spaniard, probably since he commonly had no saddle, nevertheless, in certain respects he was more clever. His ingeniousness may be explained by two principal factors: first, a saddle was lacking; and second, the hands were seldom used to guide a trained horse. The fighting native needed both hands to shoot his bow and arrow, and as a result he became a marvelous rider. Actions such as dropping on the far side of a galloping horse and shooting under the neck at the enemy, mounting and dismounting at a dead run, and picking up a fallen comrade without stopping the horse were all common feats. When it is considered that the Indians accomplished these feats without any saddle to support them and without holding the reins, guiding the horse entirely with the knees, the extraordinary skill of the Indian riders can be appreciated.

The Indians seldom had saddles unless they lived near white settlements. They rode, therefore, on a pelt cinched around the horse. Their bridles were just a thong looped around the lower jaw of the horse. The Plains Indian almost invariably kept a neck rope dragging when he was riding, so that if he were unseated, he could grab the dragging rope and retain his horse. A man afoot on the plains was as good as dead.

The immense effect this animal had on the life and habits of the Indians has never been adequately told. From a poor sedentary group they became independent nomadic tribes and woe to man or beast that got in the way of the previously humble Indian.

Comancheros

The Indians and the Spaniards of New Mexico were usually at peace. It was to their mutual benefit to trade with each other. The Comancheros, a rude and indigent class of frontiersmen, banded together several times a year and set out upon the plains with trinkets which they bartered to the savages for horses and mules. The entire stock of each trader rarely exceeded twenty dollars in value, and he was lucky if the Indians did not steal back his stock before he reached home. Considering the number of Comancheros, the total amount of their sales was considerable.

Josiah Gregg gives an interesting account of buying some horses from the Indians. He says that he and his party succeeded in purchasing several animals which cost them between ten and twenty dollars in trade goods. The main trouble in selling to the Indians lay in fixing the price of the first animal. This being settled by the chiefs, it often happened that animal after animal was led up and the price received without further argument. Each owner generally wished a mixed assortment of goods; therefore, the payment had to include several items such as a blanket, a looking-glass, an awl, a flint, a little tobacco, vermillion beads, and so on. In this manner most of the bartering among the Indians was carried on.

Texas was the favorite trading ground for the Comancheros. When Captain R. B. Marcy laid off the southern Santa Fé Trail from Fort Smith, Arkansas, to Santa Fé in 1849, he noted the camp sites of the traders along the south bank of the Canadian, their trails leading to the settlements far to the west. When Charles Goodnight arrived in the Panhandle of Texas from Colorado in 1876, there were at least three Comanchero trails almost "as plain as the wagon roads of today." The most southerly trail left the Pecos near Bosque Redondo and pointed east and south to the Yellow Houses, touching at the Plains Lakes on the way and terminating in Cañon del Rescate, in the neighborhood of Lubbock. The upper trail left Las Vegas and led northwest to the Canadian, followed down that stream to the east of Tucumcari Mountain, and forked near the Texas line. Some trading took place at the site of Old Tascosa, and some on the Mulberry in the JA range, but the most important commerce shifted to Las Tecovas, some springs northwest of the site of Amarillo, near the location of the old Frying Pan Ranch.

George Wilkins Kendall, in telling of horse buying in Houston in the early eighteen hundreds, says that anyone who entered the Houston horse market with the intention of purchasing was well aware that it was easy enough to buy a nag, but not easy to get a satisfactory one. When it was known that a man wished to buy a horse, all sorts were trotted out by the different dealers. There was the heavy American horse whose owner had probably entered

Texas by the Red River route and wished to leave by way of New Orleans, hence having no further use for the animal. There were also wiry-looking Indian ponies, which had undoubtedly been misused, the slender but game Mexican horses, and lastly, the recently caught, restless, and apparently vicious mustangs. All these horses were offered for the buyer's inspection, with the usual discussion of their merits. From so many he had no little trouble making a selection. Kendall "looked with an eye of fondness on a beautiful horse, half-Spanish and half wild, of fine action and most delicate points."

The Spanish horses were widely distributed by the Indians, French, Anglo-Americans, and Spaniards. Even before the days of the "long drives," Spanish cow horses were well known within the present boundaries of the United States, where they were spread by the constant trading, legal and illegal.

Zuñi and Apache

Although the Zuñi Indians were bitter enemies of the Navajos, they had a curious affiliation with the Apaches, at least according to Joe Mendivil, a Mexican who was captured and adopted by the Apaches in the nineteenth century. His narrative was printed in the *Overland Monthly* in 1871. Apparently the Apaches would make a grand ride to the Zuñis to trade and talk and occasionally to pick up a wife. It was a major event for the Apaches and one looked forward to with great expectation.

The horses were fed and brushed in advance, until they were smooth and glossy. They were put into a period of training for the visit, with daily workouts so that they might be among the first to reach the Zuñi village. Each Apache tried to impress his Zuñi friends with the quality, fleetness, and strength of his horse, and the splendor of his trappings, as well as with the beauty of the presents which he carried with him. Those most fortunate might have an ornate

Mexican saddle obtained in one of their forays and a leather bridle with silver ornaments. As the horses were shod in rawhide boots, many extra pairs were assembled that the horse might not become lame before the trip was complete. Often half a dozen horses were put in training before a suitable one was found.

The Apache also dressed himself as elaborately as possible. The usual loincloth was swapped for a gaudy ensemble accumulated on various warpaths. He would prefer to have Mexican garb, as the pants were open at the bottom and garnished with silver bells and lined in bright silk. Most of the Apaches, however, had to be content with their gaudy paint and a few odd pieces of clothing, such as a soldier's shirt or a trapper's hat, with perhaps some Mexican spurs to complete the effect.

All of the time prior to the expedition was occupied with a discussion of what they would do and get while on the visit. Even the immigrant trains and the glory of the warpath were forgotten in the discussion of the presents they would give and receive and the barter they would carry on.

When the dawn for the departure arrived, everything was in a hubbub, as all the village would arise to see the warriors off. They took food with them but nothing to drink. They drove extra horses with them both as presents and for trade. The gifts were for the most part gathered during their war trips and consisted of articles for which they had no personal use. Such items as fine *sarapes* from Saltillo, lariats of extra-fine rawhide, daggers, saddles which they seldom used, in short all of the remarkable and luxurious things which their village-dwelling friends might want, but for which they themselves cared little.

Finally, when they were painted beyond recognition, they left at a gallop amid shouts and cheers. As soon as they were out of sight, they slowed down to an easy pace and made most of the pilgrimage in a leisurely manner, resting wherever there was good grass for their horses.

Six or seven days were spent on the trip, which was generally about two hundred or three hundred miles. When within a few miles of the Zuñi villages, the Apaches made a final halt. The horses were fed and groomed, the paint was applied profusely to horses and savages, the presents were made readily available, and then they were ready for the full dress charge, a grand entry, a barbaric cavalcade.

It must have made an odd and picturesque panorama as the large group of savages came flying over the hills and across the plains with their plumes and feathers, *sarapes* and spurs, their gaudy trappings and colorful horses. The Zuñis, too, when they saw the Apaches coming, must have presented a sight as they crowded to the edge of their lofty, tiered adobe village.

After an exchange of greetings, everyone sat down to a large meal. When that was over, the Apaches gave their gifts to the Zuñis. Then a day or so was spent in trading and in the exchange of stories, the Apaches wandering around the village at will.

When at last the time arrived for the return, all were eager and enthusiastic to go back to their families. They had come slowly so that their horses would look good to the Zuñis and also to keep them in readiness for the homeward trip. The return was made in the fastest possible time. They planned to make it in two days and two nights. It was a race in which no quarter was asked or given. When they neared their homes, a group of riders came out to meet them and gave them fresh horses for the last dash. The next day was spent in telling of the trip and displaying the new acquisitions. Presents were in evidence everywhere, and if no new wives had been brought back, everyone was happy.

The White Steed of the Prairies

Although the mustangs now have gone to a place where the grass is always green, they have left America with an immortal horse legend. While the wild horses were roving the American plains, there arose the story of a heroic horse. Perhaps the Indians, trappers, and mustangers were responsible for the tale. After all, is it surprising that in a country where there were so many wild horses there should arise a group of songs and legends recalling the beauty and character of one particular horse? These tales are doubtless partly true, inspired by an unusual horse, and partly myth, springing from the facile minds of men who spent much of their time alone. Since many of the tales are similar, one is inclined to suspect that there is more truth than is at first apparent. The horse was generally a stallion, firmly made and supposedly more beautiful than any mortal horse, "fleet as the morning light and as elusive as the evening breeze." One time he is the color of beaten copper, again as black as a moonless night, but the stallion most often praised is the milk-white pacer with coal-black ears. This legendary figure has had many names. In one place he is referred to as the "Ghost Horse of the Plains," again as "Pacing White Stallion," in another story as the "White Sultan," in still another as the "Phantom White Horse," but more often than any other as the "White Steed of the Prairies."

All the early travelers of the plains heard of this celebrated horse. When spurs clicked and campfires flickered in the unfenced world that was the West, sooner or later someone would bring up the story of the "Phantom White Horse." And then while the campfire coals darkened, or glowed at the will of some stray breeze, all would listen to the latest feats of the milk-white stallion. News of his whereabouts was avidly sought, and for fifty years it was every youth's dream to capture and tame the "White Steed" for his own. The mustangers, a group of wild-horse traders, tried every trick in their varied and ingenious repertory. They tried snaring him, creasing him, running him down, walking him down, roping him, penning him, cornering him in a box canyon, keeping him from water, and any of a thousand and one other sure-fire wild-horse-catching schemes. But all in vain. He was apparently never caught.

Newspapers followed his movements; novelists utilized his character; poets immortalized him; playwrights eulogized him; preachers moralized on him; musicians wrote suites to him; while his would-be captors swore at him both in admiration and exasperation.

Herman Melville sketches him with unusual poetic skill in "Moby Dick." He says in part:

> . . . the White Steed of the Prairies, a magnificent milk-white charger, large-eyed, smallheaded, buff-chested, with the dignity of a thousand monarchs whose pasture was fenced by the Rocky Mountains and the Alleghenies. . . . . The flashing cascade of his mane, the curving comet of his tail, invested him with housing more resplendent than gold.

Josiah Gregg, traveling over the western plains in 1830, mentions the White Steed in his *Commerce of the Prairies:*

> The beauty of the mustang is proverbial, one in particular has been celebrated by hunters of which marvelous stories are told. He has been represented as a medium-sized stallion of perfect symmetry, milk white —save a pair of black ears—a natural "pacer." But I infer this story is somewhat mythical from the difficulty one finds in fixing the abiding place of this equine hero. He is familiarly known by common report, all over the great prairies. The trapper celebrates him in the vicinity of the Rocky Mountains; the hunter on the Arkansas, or in the midst of the plains, while others have him pacing at the rate of half a mile a minute on the borders of Texas. It is hardly surprising that a creature of such an ubiquitary existence should never have been caught.

Kendall, in his account of the Santa Fé expedition, undertaken some ten years after Gregg's journey, elaborates:

> Many stories were told in camp, by some of the older hunters of a large white horse that had been seen in the vicinity of Cross Timbers and near the Red River. That many of these stories, like the majority of those told by gossiping campaigners, were either apocryphal or marvelously garnished, I have little doubt, but that such a horse has been seen, and

that he possesses wonderful speed and great powers of endurance, there is no reason to disbelieve. As the camp stories run, he has never been known to gallop or trot, but paces faster than any horse that has been sent out after him can run; and so game and untiring is the "White Steed of the Prairies," for he is well known to hunters and trappers by that name, that he has tired down no less than three race nags, sent expressly to catch him, with a Mexican rider well trained to the business of taking wild horses. The latter had nothing but a *lazo* with him. Although he changed horses whenever one was tired, he never got close enough to rope the steed, or even drive him into a regular gallop. Some hunters even go so far as to say that the white steed has been known to pace his mile in less than two minutes, and he could keep up the pace until he had tired all pursuit. Large sums have been offered for his capture, and the attempt frequently made, but he still roams his prairies in freedom, solitary and alone.

Washington Irving mentions this white horse in his *Tour of the Prairies* (1832). He says that in the vicinity of the Cross Timbers his party had seen some wild horses which gave a "turn to the conversation of the camp for the evening. There were several ancedotes told of a famous grey horse that had ranged the prairies of this neighborhood for six or seven years, setting at nought any attempt of the hunters to capture him. They say he can pace and rack (or amble) faster than the fleetest horse can run."

Nor was it only the white man who told stories of the White Mustang. The Indians had similar legends. They were more superstitious than their white brothers. The Kiowas felt that arrows or rifle balls could not touch the "Phantom Mustang" and that he could run unscathed through a prairie fire. The Blackfeet believed that he was "big medicine" and could, if he chose, breed war horses that made the rider invulnerable in battle.

Most of the reports agree that he roved solitary and alone. Why is not generally told, although one old hunter said the reason was that he was too proud to be seen with any of his kind, being so superior in form and action to all of his brothers. It seems that many times beautiful and exceptionally developed mustangs did range alone. Perhaps they had discovered a certain safety in the lack of numbers.

On the eighteenth of May in 1892, some New Mexicans caught a horse answering in many respects the description of the White Steed. This animal was captured after a terrific chase, somewhere between the head waters of the Trinity and the Brazos Rivers. He lived but a short time because of the excessive fatigue of the chase.

It is not surprising that if this horse was caught, he was taken by the New Mexican *vaqueros*. They were one of the three greatest horse people of the seventeenth and eighteenth centuries, their only competitors being the Arabs and the Comanches. So devoted were the *vaqueros* of *Nuevo México* to equestrian exercise that they were termed "a race of centaurs" by contemporary observers.

The last report of the Pacing White Stallion comes in 1881, when he was supposed to be in the Snake country. From this date on, the legend seems to die out, and little is heard or read about him except in western novels. His span of life, although long for any mortal horse, was short for a legendary hero, since it appeared, reached its zenith, and died, all within a span of fifty-six years. There is little reason to disbelieve that some beautiful grey or albino mustang was the source of the legend, and once it started, the "White Steed of the Prairies" became every man's dream horse.

8. Recreation Became a Business

☙ *Soldados de Cuera*

Charles III of Spain decided in 1768 that the northwestern portion of his empire in North America should be settled. Earlier Spanish voyages had already given him knowledge of the coast line, so that when Russia seemed to constitute a threat, he ordered Francisco de Croix, viceroy of New Spain, to settle California. San Diego and Monterey, two spots known to the Spaniards, were chosen as first objectives.

In the Spanish Empire oftentimes orders of this nature were not fulfilled with the celerity that met this suggestion. In New Spain at that time was one José Gálvez, who not only was a dynamic personality but was clothed with all the power and means to complete an undertaking such as that proposed. To cinch the occupation, he decided to send two expeditions by sea and two by land; surely then the best route would be discovered and the establishment made. It was an insurance against failure of any particular group.

The expeditions by boat interest us little, but the land expeditions carried the first Spanish horses into California, the land where they were to increase and multiply until they became absolute pests like the rabbits of Australia. The overland expeditions were under the command of Fernando de Rivera y Moncada and Gaspar de Portolá. Military security was obtained when Pedro Fages was assigned to accompany the California venture with a small company of Catalán volunteers.

Supplies for the California venture were to be obtained from the missions of Lower California, which by this time were already established. Even the livestock were to be obtained in this way. José Gálvez, wishing to have the expeditions start while he was in that area, could not wait for the livestock to be driven overland from Sonora, although Sonora, established longer and in a more ideal livestock country, would have been the logical place to obtain the stock. Later, of course, it did furnish most of the animals so that the Lower California missions would not suffer unduly; and the original order specified that the livestock taken from Lower California were to be replaced by animals from Sonora at the earliest possible moment.

Don Fernando de Rivera was commissioned to obtain the animals for the overland expedition. Don Fernando was *capitán* of the *Soldados de Cuera,* and second in command of the land expeditions. The *Soldados de Cuera* are worth an added word. Literally translated, *soldados de cuera* means "soldiers of leather," and so they were. The Spaniards early learned to discard their heavy metal armor in the hot lands, for metal in a desert sun is not the most desirable type of clothing. Also, the difficulty of repair or replacement on the frontier is obvious. Therefore, it rapidly became neither practical nor acceptable, and the Spaniards made a very suitable armor from hides. They fashioned them into long leather jackets, without sleeves. Several thicknesses were used and the leather was quilted to cushion the blows and to lessen penetration. On their legs they wore *chaparreras* similar to the American "stovepipe" chaps. They also had guards or aprons of leather which they tied to the saddle horn and placed around the breast of the horse to protect the animal as well as their own legs. This type of armor would turn Indian arrows except at very close range. Later the same device was adopted with minor variations by the *vaqueros* as a protection against brush and cactus.

Rivera, beginning in the south, worked north through Lower

California, gathering supplies, including horses and mules, for the venture. He started with the royal presidio and mission of Loreto, where he got a good start, and from there he went to San Francisco Xavier. Here he obtained sixteen broken mules, which could also be used for riding if needed, and four horses. From San José de Cummundía he obtained twenty-six more mules, which could be used for either packing or riding, and six horses. From Purísima de Cadigomo he obtained seven mules and four horses, two actually with saddles and bridles. From the mission of Guadalupe he obtained sixteen broken mules, four good horses, and ten sets of harness. From Santa Rosalia, fifteen mules and three horses were obtained, and from San Ignacio, twenty mules, six horses, and two jennies. From Santa Gertrudis, three broken mules and four good horses were added. From San Francisco de Borja, sixteen mules, eight horses, six mares, one stallion, ten sets of harness, two hundred head of cattle, a hammer, and some branding irons were secured. From Santa María de Los Angeles, four mules were obtained. In this manner Rivera gathered stock from the missions and prepared to move them into Upper California.

The above list, however, does not represent by any means all the livestock going to California. It was merely the equipment thought necessary for one of the original expeditions. Two years later arrangements were made with the Viceroy to stock more adequately the Upper California missions. For example, we have the following livestock ordered sent to San Diego by the Viceroy:

From the mission of San Borja, 150 cows with their calves, 25 bulls, 40 fillies, 2 stallions, 6 colts, 1 jack for breeding, 50 ewes and 16 rams, 200 nanny goats and 20 billy goats.

From Santa Gertrudis, 30 mares and 1 stallion and a jack for breeding, 5 colts and 6 fillies, 50 ewes, 16 rams, 100 nanny goats and 10 billies.

From San Ignacio, 16 fillies, 5 horse colts, 1 stallion, 4 jennies, 2 jacks, 125 ewes, 25 rams, 50 nannies and 6 billies.

From Guadalupe, 1 stallion, 6 jennies, 3 jacks, 100 ewes, 40 rams, 80 nanny goats and 10 billies.

At the foot of this order are found the following words: "The hogs, I believe, will go best by boat, and for that reason I make no mention of them here."

◆§ *The California Grizzly*

When Rivera had finished gathering the necessary articles at Mission Santa María for his expedition to California, he wrote José Gálvez, the man who had directed the formation of the California venture, that he was ready to start. He also wrote Junípero Serra, who was to head the mission, that he was ready to start for San Diego in March.

Rivera soon found that Mission Santa María did not offer sufficient pasturage for all his livestock. He had at this time about five hundred head of horses, mules, and cattle, so he moved on to Velicatá. As Father Serra could not go with the original expedition, he sent Father Crespi, and the expedition got under way. Portolá, with the second land expedition, was already on his way to Velicatá from the northern frontier of Mexico. He left Velicatá, in company with Father Serra, not quite two months after Rivera, and followed the tracks of the preceding expedition. Both arrived in San Diego with little major difficulty.

In San Diego, Father Serra's old foot injury began troubling him again. The long trip had not helped it. The affliction was so serious it seemed as if he might not be able to continue. Calling one of the *arrieros,* or mule-skinners, who was in charge of the livestock —one Juan Antonio Coronel—he said, "Son, canst thou not make a remedy for the ulcer on my foot and leg?"

Juan answered, "Am I a surgeon? I am a poor *arriero,* and I have healed only the sores of animals."

"Then, son," replied Serra, "suppose me a beast and this ulcer a saddle gall from which has resulted the swelling of the leg and the pains that I feel and that give me no rest. Make for me the same medicament that thou wouldst apply to a beast."

Palou records that Serra afterwards said, "Now the foot itself is all sound like the other, while from the ankle half way up the leg, it is as the foot was before, an ulcer, but without swelling or pain, except occasional itching. In fact, it is nothing serious." Apparently the amateur veterinarian did a fair job. At least Serra was soon able to travel again, although he was to have more trouble with his leg.

After a suitable time, Rivera and Portolá decided to push on from San Diego in order to settle Monterey as they had been ordered. They found the journey rather difficult, with water and food very scarce.

Near the present city of San Luis Obispo, they came to a small cañon which came to be called *Cañada de los Osos* because there they encountered a group of bears. The soldiers decided to attack them immediately. They needed fresh meat and, at least, would have great sport. Although they succeeded in killing one bear with their muskets, they had their first taste of the ferocity of the grizzly. As soon as one of the bears was struck with a shot, he immediately attacked the nearest horseman at full speed. When this occurred, the Spaniard needed the full dexterity of his horse in order to evade the charge. A wounded bear never gave up, and he would not turn and run. The only time he stopped fighting was when he fell dead. It soon became apparent that the only places to shoot bears were the head or the heart, for only then did they die instantly. The Spaniards foolish enough to pursue the few unhurt animals which had lumbered away after the first shots were lucky to escape with their lives, especially if their horses were not exceptionally good animals. One bear upset two horses, and it was only by extreme good fortune and the grace of *el buen Dios* that the episode was not more costly.

A little farther north the men again ran out of food. They became so hungry that they lived for a while on sea gulls and mule meat. It is recorded that only the Lower Californians and the Catalâns had any stomach for the mules.

◆§ *Vaqueros*

As each California mission and presidio was established, horses and cattle were supplied from the sister missions in the south. By 1784 there were enough cattle to feed the settlers, and by 1800, with the help of the California climate and natural grasses, the cattle had increased so much that never again until the Gold Rush was California to know want in these animals.

Cattle were almost from the first easy to obtain. This is clearly indicated by their value. They were listed at ten dollars in 1781. By 1800 there were in California 74,000 cattle, 24,000 horses and 88,000 sheep, showing the rapidity with which the original animals reproduced in the new environment. Most of these, of course, belonged to the missions, but secularization, which was now taking place, led to wide diffusion of the animals, many living in a semiferal state until the arrival of the Anglo-Americans.

The cattle were leggy and tough, with long horns, brothers of the famous Texas "longhorn"; and the horses were small and wiry, although not quite so small as the average horse in the Southwest, probably because of the more abundant feed and adequate minerals available in California. New blood was brought in, and the old breed rapidly became extinct. Cattle was slaughtered for food in a most wasteful manner by the goldseekers. Wild horses ate up so much grass that thousands were driven off the cliffs at Santa Barbara, into the straits at Carquinez, and into corrals elsewhere, to be slaughtered. By 1850, with the influx of the Americans, cattle were worth from three hundred to five hundred dollars each, although soon thereafter, when the newcomers began to raise livestock to supply the market, the price dropped accordingly. Several herds were driven all the way from Texas, where rumors of the fabulous prices paid for cattle had been received. Before many years, the livestock industry revived; and California, in turn, began supplying the neighboring areas with their foundation herds, particularly those of the Northwest. The re-establishment of the old industry, once started, was rapid. The census figure of 262,000 cattle in the eighteen fifties rose within a decade to 1,180,000, and three years later, in 1862, was 3,000,000. Except during several severe droughts, cattle raising has since that time continued as one of the leading industries of California, and the horses, an inseparable part of that business, have kept their place.

Each mission had several *vaqueros* (cowboys) who were generally Indians from the *ranchería.* The *vaqueros* performed all of the tasks common to cowboys, under the supervision of the *mayordomos* and *caporales.* The officials and certain few *vaqueros* rode saddles; most of the other men used none. Those who rode saddled horses were furnished with all the necessary equipment—saddles, bridles, ropes, and spurs—by the mission. The others received only the minimum equipment given all the natives. With the exception of the *vaqueros,* no other Indians were ever allowed on horseback, for since the earliest days giving an Indian a horse had always been one of the most serious offenses in the Spanish American World.

Until 1853 wheat was always separated by horses. The harvest was carried into a special corral, a yard often one hundred feet square, with a stone surface and a substantial fence. Sometimes the straw would top the fence. A group of from thirty to sixty mares was driven in upon the straw. The Indians would then start the group milling around the corral, and the grain was soon threshed.

Many horses and cattle ran in the outskirts in a wild state and caused no little trouble among the ranchers. The years between 1821 and 1824 were particularly bad, as the wild livestock ate much of the pasture needed for the domestic herds and caused many of the tame animals to join their wild relatives. The government, backed by the *rancheros,* decided to stop this encroachment. Carmen Lugo, in a manuscript preserved in the Bancroft Library, says:

> I remember having seen three corrals here in Los Angeles—two were made by the people of the town and one by my father. Many other per-

sons also made corrals for the same purpose. These corrals were erected just outside the town. After rounding up the horses, the cowboys would take them to the corral and hold them. Then they would open small gates through which only one animal at a time could pass. As the wild horses came out two or three men placed right next to the doors on the outside, killed the horses with lances. This was done, of course, after the tame animals among the wild herd were separated. They killed all of the horses which were unclaimed. I know there were thousands of wild horses killed at that time. There were many killed in another corral built for the same purpose at the Nietos ranch.

The Californians had a peculiar method of breaking wild horses which deserves notice. The colts were allowed to run wild until they were four or five years old. Before that age they might be lassoed half a dozen times, always at least once, to be branded. To be caught, a colt had to be driven into a corral with the herd and roped. The rest of the horses were then turned out. A second lasso was thrown on the colt's forefeet, and between the one on the feet and the one on the neck, he was soon thrown. A *vaquero* then sprang on his head, and holding it down, and covered his eyes, while another *vaquero* tied all the feet securely together, so that the colt could not possibly arise. A *jáquima,* or headstall, with a long hair rope and a piece of leather arranged so that it could be drawn down over the colt's eyes to blind him, was fastened on securely. The ropes were removed from his neck and feet, and he was allowed to rise. After rearing and plunging, he grew tired and, finding himself securely fastened, would quiet down. One of the *vaqueros* would then approach him very carefully and pull the blind down over his eyes. This act required a great deal of time and was often dangerous. As soon as the blind was down, the colt would stand. Then a *vaquero* would put on his saddle. After it was well cinched, the blind was raised. Again the colt would pitch and snort and often throw himself. When this excitement was over, the *vaquero* would advance and put down the blind, and the colt would stand trembling with fear and exhaustion. The *vaquero* tightened the cinch, tied the long *mecate,* or hair rope of the *jáquima,* to serve as reins, fastened the colt's ears under the halter so that he could not hear, threw a leather strap over the saddle, and mounted. He threw his knees forward, bent at right angles, and tightened the leather strap over them, so that he was really tied on.

When he had loosened the colt's ears and raised the blind from his eyes, he was ready to ride. The colt now gave his major demonstration. The *vaquero,* practically tied on the saddle, rode easily. The stiff-legged jumps were hard to take, but the excitement repaid the work. Within a short time the horse would stop pitching and be content to run. The rider then put down the blind, dismounted, removed the saddle, raised the blind, and let the colt rest until the next day. If he had bucked violently and promised to be very troublesome, he was tied with his head up so that it would get stiff and the colt would not be able to buck so viciously the following day. He was ridden every day until broken. The hackamore was always used until the horse was completely broken; this kept his mouth in good condition and he learned to obey the reins of the hackamore as easily as he afterwards would the bit. When the bit was first placed on the California horse, he would sometimes pitch, although perfectly peaceful with a hackamore. To make stiff-legged jumps seemed a characteristic of a California horse.

The use of the hackamore instead of the bridle has its good points. One of the dangers of starting with a bit when breaking a horse is that the horse may rear and fall back. When a horse is started with a hackamore, he seldom does this. A hackamore also protects a young horse's mouth, and was especially advisable in California where spade bits were used.

The strap over the saddle for breaking colts was of great service. It made the rider part of the horse, as the horse could not make a motion without the rider sharing it. At the same time, the rider, by straightening his legs, could free himself instantly in an emergency.

◆§ *Fun Astride*

Recreation was the Californian's business. Indeed, to pass time pleasantly was to the Californian the most serious consideration in his world, comparable to eating, drinking, and religion. The last was possibly thrown in to insure the perpetuation of this ideal state. Since California men's lives were so closely tied up with their horses, it is scarcely surprising to find that most of their sports included, and were executed on, horses. A description of their games also reveals the equine sport of all Spanish America.

One of the favorite pastimes was a bear-and-bull fight. A holiday was sufficient pretext, and holidays often arrived. *Vaqueros* would be sent out to rope and bring in a bear to the town or ranch which was to have the sport. Today it would seem quite a trick to find and rope a grizzly bear, but there are too many authenticated documents attesting this feat to doubt the accuracy of the statements.

One California rancher had started over the hills one day to visit a neighbor when he saw a bear. Taking down his reata, he roped the bear which, contrary to expectation, the moment he felt the rawhide rushed the horse and rider. Being unable to tighten the reata, the man had no alternative but to flee. Not wanting to lose his good rawhide reata, he did some quick thinking. Passing under an oak tree, he threw the end of his reata over a limb and, catching it, took his dallies around the horn and in less time that it takes to tell, had the bear hanging from the tree. Taking a couple of turns around the trunk, he had the bear fast and was soon again on his way.

The normal procedure for obtaining bears is interesting. The *vaqueros* traveled in groups, and when a bear was found, one man roped it by the head and another by the hind legs. The horses then stretched the bear out, and other riders would rope and tie him so that he could be put into a wagon.

When the bear was obtained and the day for the function ar-

Mexicans roping a grizzly bear. (From *Pictorial California*, by Edward Vischer [San Francisco, 1870]; courtesy Bancroft Library, University of California)

A California bear-and-bull fight. (From *Pictorial California*, by Edward Vischer [San Francisco, 1870]; courtesy Bancroft Library, University of California)

rived, a bull was also brought to the ground. The two animals were then tied together, with a rope somewhat less than thirty feet long, depending on the size of the ring or plaza. In the subsequent trial sometimes the bear won, and sometimes the bull. Once at a fight in Santa Barbara a bear killed three bulls in succession. A bull would start the fight, and the bear merely defended himself, and ended it. On another occasion a bull killed a bear with a single thrust of its horns.

The fight generally took place inside a strong wooden fence, behind which, at a short distance, was a platform for the women and children. The men watched on horseback with ropes ready in case of need. In Monterey, California, there is still to be seen the remains of a bull and bear ring made from adobe and wood.

Bullfights were also held, although normally the *golpe de gracia,* or death blow, was not given unless some visiting toreador from Mexico City or Spain was present. Most of the towns had plazas built with bullfights in mind. The bull was brought in, roped by the horses, and then turned loose. Various local ranchers would try their skill at sticking the colorful darts between the bull's shoulders as the beast thundered by, or they would tease it with their *sarapes* to the enjoyment of the crowd. When the bull was tired and everyone wishing to enter the arena with him had had a chance, the gate was opened and he was turned out.

At this point all the men who had been waiting on horseback had their opportunity. They would *colear* the bull as he ran out. This was called the *corrida de toros* and was a particularly popular sport, as many could enter the game at the same time. As the bull left the arena, all the horsemen would start after him pell-mell and try to grab him by the tail and throw him. With so many horses and men jostling about, the game proved almost as dangerous as it was popular.

Men would often *colear,* even when there was no *corrida,* or bullfight. Don Mariano Guadalupe Vallejo, one-time governor of

Colear—a Mexican *charro* throwing a steer by the tail. (From a photograph in the author's possession)

California, said he used to *colear,* but that his brothers, Juan Antonio and Salvador, were so skillful that one time they had their *vaqueros* line up six bulls for each, each bull about 250 yards from another, and then they rode to see which could throw his six first. That could be considered "throwing the bull" creditably, even today.

Horse racing was always a favorite pastime, but when combined with *Juego de Gallo,* it was perhaps the top sport. In *Juego de Gallo,* a chicken was buried with only its head exposed, then a rider would dash by at full speed, trying to swing down to grasp the chicken by the neck and come to the end of the course with the head in his hand. As the chicken could and would move its head, this called for real skill and horsemanship.

Juego de Vara was another favorite and is somewhat reminiscent of several modern games. All the men who wished to participate in the game formed a circle facing the inside. Room enough was given between each horseman to allow him to turn around. A rider carrying a flexible wand rode around the circle slowly until he decided to give the stick to someone. The receiver immediately wheeled and gave chase, trying to strike the rider as many times as he could before he had made the complete circle and entered the vacant spot left by his pursuer. The same game is still played by schoolboys today on foot and generally with a roll of newspapers or a belt strap.

The early Californians even had fun from their work. A good example is *nuquear.* This was done on horseback and was considered the most satisfactory way to kill cattle, since it was sport. Normally, whenever a ranch owner wished to slaughter, he sent six riders on good horses to be the *nuqueadores.* With swordlike knives in hand, these men rode at full speed after the animal which was to be killed and skillfully chopped it on the neck between certain vertebra, severing the spinal cord. The animal would fall dead instantly when struck correctly, while the *nuqueadores* kept going

Carrera de gallo, "snatching the rooster." (From *Hutchins' California Magazine* [San Francisco, July, 1860])

ahead *nuqueando* to the right and left as they rode. If they missed the correct spot, nothing happened. It required skill, and the competition to see who could kill the most animals added the necesary zest to the work.

All the Californians were great hands with a rope, and they were always amusing themselves roping objects and one another. Alexander Forbes, who wrote a history of California in 1839, said that the first object one saw in a little urchin's hand was a lasso of twine with which he essayed to ensnare his mother's chickens. Roping contests and games involving the reata were common. Stories are told by early writers of certain individuals even roping low-flying geese. So skillful were the cowboys that at the rodeo they could rope any one or all of the feet of the animal desired. They used rawhide reatas, often as long as sixty feet. They never tied the rope to the saddle horn as is common in the Southwest but took the so-called dallies around the saddle horn to hold the rope after the animal was caught. The word "dally" comes from *da vuelta,* meaning "to take a turn," just as lariat comes from *la riata,* mustang from *mesteño,* and hackamore from *jáquima.*

◆§ *Montadura*

Carriages and automobiles were unknown in pastoral California, but the inhabitants never walked. It is not surprising, therefore, that the *caballero,* with his natural love of splendor, spent his money for *montadura,* or riding equipment. And why not? There was no other use for a thing as plentiful as the cattle on the hills and as little valued. So in the golden age we have a silver age, when the silverwork and the saddle art reached a new peak of excellence. Engraved leather-work also flourished, and the skill of the pastoral Californians is considered by some experts to be of the best.

Leather-stamping is of Moorish origin and dates back to the Moorish invasions of the eighth century. From Spain it was brought to Mexico and taught to the natives, who soon exceeded their teachers in ability. The Aztecs were already metal workers and they were expert sculptors and engravers. Soon they were making beautiful equestrian equipment. When the Spaniards came to California (where the affairs of the nation seemed insignificant when compared with the importance of the *caballero's* trappings), it is small wonder that the art blossomed afresh and reached a high peak. Today there seems to be a revival of the old yen for beautiful *montadura.* Traces of the old Spanish skill were never completely erased, although for a time they seemed but a memory. When the great influx of the Americans occurred in 1849, most of the Latin ranchers, with their cattle and horse herds, disappeared; and the wealth passed into the hands of the shrewd, money-loving Yankees, against whom the hospitable, free-hearted Californians were no match.

Today if you go to Santa Barbara around *fiesta* time, or any of a number of California towns when they are holding their various pageants and rodeos, you will see something very similar to the old California. You will see mounted *vaqueros* with braided *reatas,* and even bridles and reins exquisitely plaited of rawhide, silver, or a combination of the two. Silver saddlery in the early California style is again popular.

The California bit was a massive piece of metal, elaborately trimmed and inlaid with burnished silver, delicately wrought, and attached to the bridle with chains, usually of silver. The mouthpiece consisted of a spade or ring bit, sometimes even it was made of silver.

The ease of the old Californian on horseback is proverbial. No three-legged riders they. Flat in the saddle they sat, and with their light hand the bit lay nicely in the mouth and was never jerked to turn or stop. Changes in pace or direction were given to the horse by shifting the weight of the body in the stirrups or the touch of the rein on the neck. Unless the horse was vicious, he seldom felt the bit. Also, the horses were all broken with a *jáquima,* or hackamore, the bits coming after the horse was broken. This practice is still followed in California. Properly used, the bits assure perfect control and are not severe. Californians always ride with a loose rein, and to do so properly is indeed an art. The ability to ride with a loose rein is appreciated by good riders—and horses.

Small beadlike metallic cylinders line the bit from the side to the spade that goes well back into the mouth, and in the mouthpiece is inserted a small fluted wheel. If the question is asked why this small roller is used, the usual answer is that the horse likes it—probably as good an answer as can be given. One might wonder what developed the taste and how the owner knew about it. The horse will invariably roll the little wheel with his tongue when not otherwise employed, for his amusement, apparently, so maybe he does like it. When the rollers are omitted, the horse is commonly said to be sour bitted. Some Californians swear his good nature suffers when the wheel is not present. Occasionally the ring bit was used. It has come to be called a "jaw-breaker." This bit does not project as far into the mouth and encircles the lower jaw like a chin strap, thereby giving an immense leverage to apply when desired.

A riding whip is not a California adjunct; a long leash attached to the reins, termed a *romal,* takes its place, although the spur relieves it of much active work. Some of the old ornamental spurs had rowels six inches across, although today from one-half to three-inch rowels are used, and they are blunt. These, when provided with chains and jingling appendages, added much to the color and tone of the *caballero* and his *montadura.* When a halter rope was used, it was generally firmly plaited hair from mare's tails, strong and durable, with often two or more colors beautifully blended and the tip furnished with an attractive tassel. The hair of mares was used because it was considered an insult to cut a stallion's mane or tail.

Hubert Howe Bancroft in his *California Pastoral* tells a good story about the reluctance of the Californians to tamper with the mane or tail of a stallion. At a dance held in California during the early days a practical joker stole out to the hitching racks and cut all of the tails of the riding horses. The Californian, of course, would not ride a mare; that was much too effeminate. When the dance was over and the men saw the dastardly trick which had been played on them, they were furious and looked for the culprit in order to punish him summarily. However, the joker had been clever enough to cut off his own horse's tail so that he was never caught. The affair caused a sensation and was long remembered.

In the New World the Spanish blood was crossed with native American stock, and here we have a people already highly artistic receiving a new infusion from a race equally sensitive to color. Where else but in Spanish America could you have an ordinary situation which boasted of a woman in a house of two rooms with a mud floor "dressed in spangled satin shoes, silk gown, high comb, and gold earrings and necklace?" Arthur Chapman, some hundred years later, echoes this situation with this little jingle:

He had found life's secret,
He had traced it to its source
With his hundred-dollar saddle,
And his twenty-dollar horse.

José de Carmen Lugo, originally wealthy owner of the Rancho San Antonio, left a manuscript in 1877 entitled *Vida y aventuras de mi padre*. From this interesting hand-written document comes much information about the attire of the Californian. A man never cut his hair. It was parted in the middle, braided into one braid of three tresses, and then thrown over the back. A woman let the hair cover the ears, parted in the middle and braided as the man's was. The man cut his beard short, leaving sideburns from the temple to the border of the jaw *("orilla de la Quijada")*. He usually shaved every four or five days, and generally on Saturday afternoon. As a rule he bound his head with a black silk handkerchief, knotted either in front or behind. This was topped by a hat secured with a deerskin band or silk ribbon, depending upon the purse of the owner. More often than not the hat was tipped to one side or the other as the owner changed moods. The average hat was called a *poblano,* being made at Pueblo, although the rich had fine woolen *(vicuña)* or leather *(vaqueta)* hats. Occasionally a palm hat was seen, made by the Indians. *Poblanos* were curved with a wide brim.

Among the women of early California there were some wonderful riders. Some of the women rode astride without regard for convention and managed the horses like masters. Some could rope like men; such a woman was Señorita Josefa Arguello, daughter of the governor. Often the women rode alone, but on the way to a dance the men rode behind. The women occupied the saddle; the men mounted on the *anquera,* a leather covering extending behind the saddle just for such purposes, and guided the horses. One advantage of this method was that the arms of the man encircled the fair horsewoman.

Some of the early Americans in the country realized that the old California was rapidly giving way to a new, and the thought of preserving the art and the tradition of the older inhabitants was born. To Mr. D. W. Thompson of Santa Barbara, a forty-niner, goes most of the credit for originating the plan that today gives Santa Barbara at certain seasons an almost barbaric splendor in the horse trappings and the traditional garb of the Californian of the golden age.

To preserve the *montadura,* he had a saddle and bridle of great beauty made entirely of Mexican leather, stamped in the incomparable style of the time. Mexican silver dollars were collected and given to silverworkers; the saddle was profusely decorated, and the bridle was heavily laden, with the silver obtained from them. Since the Mexican silver dollars contained less alloy that those of the United States, they worked into shape better. Each part of the saddle was bordered with rows of silver rosettes, the pommel encased in straight silver, and the cantle and stirrups were almost solid inlay.

Some of the ornaments were flowered; others had wheat heads in laminated silver most delicately formed and chased. Nothing but silver entered into the construction of the reins, throatlatch, romal, and martingale. The dollars were cold-drawn into fine wire, crocheted into sections joined with solid links and rings. The bridle was covered with fluted silver, with the exception of the crossed brow band, which consisted of two slender chains which crossed the face under a six-pointed star. The brow band and nose piece were finely engraved. The bit was the best that could be made. The bridle with its attachments weighed twelve pounds and more than 250 silver dollars were used just in its construction. When Thompson appeared in any city in California with his favorite "Canute" decked out in this famous *montadura* with the glittering equipage reflecting the sunlight from a thousand points, it was a sight to behold; and once seen, it could never be forgotten.

Today Dwight Murphy of Santa Barbara is helping continue the tradition. He is preserving the tradition of the early California not only in respect to the *montadura,* but also in the breeding of Palomino horses. Sheriff Eugene Biscailuz of Los Angeles and Jack Mitchell are two men who are continuing the California heritage.

9. The Gringo Accepts the Horse

Philip Nolan

The trade in horseflesh carried on by Spaniards, French, Anglo-Americans, and Indians did more to spread horses than all the wild droves on the continent. Legitimate and illegitimate trade are inextricably interwoven in the traffic. Since the Spaniards always considered international trade detrimental, much of the trade was necessarily contraband.

The new American colonies, after the Revolutionary War, still received the majority of their horses indirectly from the Spaniards. Even such a well-known Virginia country gentleman as Patrick Henry gave specific orders to a friend detailing the type of saddle animal he wished brought back for his personal use from the Pawnee country across the Mississippi; and, according to some reports, Daniel Boone was not above crossing the river for Spanish horses. This contraband movement of Spanish horses reached its peak between 1750 and 1850, and the two men who stand out above the rest in this traffic are Philip Nolan and Peg-Leg Smith.

The name of Philip Nolan is the better known because an author accidentally used his name in a popular patriotic story. The real Philip Nolan was reared by an army officer. He was accused of treason by more than one nation, and, although occasionally he may have run guns across the border, his principal contribution was his demonstration to other American frontiersmen of the comparative ease with which Spanish Louisiana and Texas could be penetrated and livestock secured. Philip Nolan, leading one group of American adventurers after another into foreign territory, laughing at the Spanish authorities, and boasting how twenty of his men could defeat a hundred Spanish troopers, added glamour to the beginning of an epoch. His eccentricities made him a legendary figure, and his striking personality gained for him friends and enemies in all classes.

Nolan may or may not have had political and military reasons for his excursions, but the important facts now are that he was a friend of the Comanches and a trader in forbidden territory, and his trips brought nothing but alarm to the Spaniards. Whether Nolan was a horse trader or a conscious exponent of Manifest Destiny makes little difference to this narrative.

Nolan made several trips to the Texas Indian country for horses. He went to San Antonio in 1795 and rounded up 250 horses which he drove back to Natchez, selling the last in Frankfort in June, 1796. Later he drove 1,000 head from San Antonio. He maintained pasture lands on the Medina and Trinity Rivers. These were used to break the long trips and provided a place where exhausted horses could recuperate before being sold across the Mississippi or in New Orleans. Nolan also had a ranch at Natchez where he wintered and broke his horses.

In October, 1800, Nolan departed for Texas on his last trip, taking with him a group of some twenty-five men. When the party was about forty miles from the Washita River (they crossed the Mississippi at Walnut Hills), fifty Spaniards on horseback were encountered. The Spaniards were looking for Nolan, but, having a healthy respect for the long rifles of the gringos, said that they were searching for Choctaw Indians.

Nolan continued to the Red River and, building a fort, crossed at Old Caddo Town. In need of fresh horses, the party traded with the Twowokanas Indians and received new mounts. Crossing the Trinity River, they camped at the famous "Painted Springs," where a treaty of peace had been recorded on a rock by the Comanches and the Pawnees. Since game was scarce, they shot some wild horses for food. Then continuing toward the Brazos, they found elk, deer, and horses. After making camp and building a corral, they successfully obtained a large number of wild horses. This camp must have been near Waco. Soon after the capture, a group of Comanches asked Nolan to a parley to be held on the South Fork of the Red River. Seeing no way to decline, Nolan accepted the invitation, and some time elapsed before he was able to return from the conference. He did not dare let any of his men leave him, knowing full well the bargaining power of his long rifles; and as a result, all the wild horses he had corralled died of thirst. Cotton fields along the banks of the Brazos are probably growing over them today.

Soon after Nolan's return to the camp, his party was attacked by the Spaniards. One of Nolan's men left a diary with an account of the fight:

> They surrounded our camp about one o'clock in the morning on the twenty-second of March, 1801. We were all alarmed by the tramping of their horses and as day broke, without speaking a word, they commenced fire. After about ten minutes, our gallant leader Nolan was slain by a musket ball which hit him in the head.

Traders from New Mexico

During the nineteenth century the people of the United States were beginning to cross the Mississippi and settle in that great area of land lying to the west. The only settlement in the Far West of any age was Santa Fé, which had been a Spanish and Mexican stronghold since the beginning of the seventeenth century. Thus it was Santa Fé that became the principal stopping place. It provided the last settlement on the way to California and the first civilization on the way back. A little later Salt Lake under the control of the Mormons furnished another axis and much the same type of assistance. In the early days livestock was taken from the Pacific coast by traders from New Mexico and the United States. Later we find the Californians also driving their stock east.

One of the earliest ventures undertaken by an Anglo-Saxon to obtain livestock on the Pacific coast and drive them east was accomplished by one Richard Campbell. When Campbell was in New Orleans in 1826, he decided that there were financial possibilities in bringing California mules to work in the Louisiana sugar industry. Therefore, in 1827, he left New Orleans and headed for California. At Santa Fé he picked up thirty men and pushed on to San Diego and, to quote, he "found no difficulty throughout the whole distance." However, other interests arose, so that he never drove any mules east from California.

A few years later, in 1830, Ewing Young was also in California with the same idea. Only his beaver exploits have been stressed in the past, but he was there frankly for two reasons: to trap beaver and to buy horses and mules. Young wrote a letter to J. B. Cooper in 1830, in which he said: "It was my intention to return from the Red River [Colorado] in December . . . and buy mules . . . but . . . I must drop all idea of the mules speculation for this year." When Young left California, he took 1,500 horses to Santa Fé.

Ewing Young intended to return to California but could not, and David Jackson took his place. Jackson, with Ewing Young and David Waldo, had organized a private company with the avowed purpose of carrying on merchandise business in New Mexico, hunting beaver, and purchasing mules in California for the sugar plantations in Louisiana. In the fall of 1831, a party under Jackson's leadership launched the expedition. The party was composed of some eleven men and, luckily for the historian, Jonathan ("Juan Largo") Warner. Warner lived a long and remarkable life, leaving behind a manuscript full of reminiscences, including an account of the venture of Jackson.

Each man was given a mount to ride and for the length of the expedition was to receive twenty-five dollars a month, not a great sum, but if the expedition required a year, Jackson's payroll would be at least three thousand dollars. This amount is an indication of the profitableness of such an expedition, for three thousand dollars was no paltry sum in those days. Seven mules were loaded with Mexican silver dollars, carried along as purchasing power.

Jackson decided on a southern route, and so moved down the Río Grande past the copper mines, then to Tucson, and on to the Colorado. After starting, he decided that it would save him money to have someone in California gather livestock for him while he was on the way. Then when he arrived, he would not have to spend wasteful months in search of animals. Jackson knew that a man by the name of Cooper was living on the coast and had been since the time of Wolfskill's party; to him Jackson wrote as follows:

> I call upon you as a friend unknown to act as an agent for me in sending up and down the coast for all the mules from three years up to eight [years of age], not to exceed 100 [in number]. The expense I will pay on sight, up to eight pesos for broken mules according to [their] quality.

The expedition succeeded in crossing the Colorado desert rather happily and made their way slowly on to San Diego, which they reached early in November, 1831. Here, to Jackson's disappointment, he did not find any mules collected, and it was necessary for him to continue as far north as San Francisco in order to buy the type of animal he felt suited for his needs. He obtained only six hundred mules in all, but did pick up one hundred horses. The number was so disappointing that he decided to sell the mules in Santa Fé instead of driving that small a number to Louisiana, as he says, "by way of Texas." The Louisiana sugar mills would have to do without California mules for now.

The adventurers mentioned above who came to the Pacific coast for livestock in the early days do not represent the entire group, but they are good examples. Governor José Antonio Chávez of New Mexico called attention to the trade in livestock between California and the East at this time. On May 14, 1830, he wrote to the minister of the interior of Mexico, telling him of the mule trade with California. An idea of the amount of horse and mule trade can be gained clearly from the following governmental command issued in California: "Jan. 1839. The traffic in mules and horses for woolen goods [from New Mexico] which has hitherto been carried on in the areas, is hereby absolutely prohibited."

The order went on to suggest that the missions get their own looms in operation so that the Indians would not have to hide their nakedness in New Mexican woolens, but might use California's own wool, which the officials (typical Californians) felt was superior. However, this regulation against the trade was about as effective as a certain prohibition issue in the twentieth century.

Peg-Leg Smith

In these early days stealing was as common as trading and almost as honest. Illegitimate horse trade became so common in the Far West that the authorities found it necessary to take special precautions. It became obvious that whenever the New Mexican traders were in California, along with certain trappers and Indians, there was trouble.

Perhaps the greatest of all horse-stealing expeditions to California set out in 1840, led jointly by Peg-Leg Smith and Chief Walkara. Peg-Leg Smith was a famous Rocky Mountain man. In one of his expeditions on the upper Sacramento River, a musket ball had lodged just below his knee in a fight with the Indians. Being hundreds of miles from relief, he was forced to amputate his own leg with no instruments other than a knife, a flint, and a hot iron. The operation was successful, and he soon was stamping about on a wooden leg. It is difficult to discover just what his true occupation was—his friends knew him as a free-lance fighter and adventurer and his enemies as a horse thief.

A more colorful figure than Peg-Leg Smith would be hard to find. He was born in Garrard County, Kentucky, and is supposed to have uttered a war-whoop at birth. As he himself said, he was "domesticated among the painted warriors of the buffalo grounds." From Fort Hall down to Albuquerque and from Independence across the plains and mountains as far as Los Angeles, the name of "Terry-oats-at-an-tuggy-bone" (choice Utah for "good friend")—as the Indians called him—was known and respected.

Peg-Leg was on the Green River during the early days of his career with a small group of men waiting for some supplies when a band of French trappers arrived accompanied by some six hundred Utahs on a "beef" hunt. Peg-Leg borrowed a horse from Sublette, one of the party, and decided to join in the sport. Sublette, keeping his best horse, gave Peg-Leg one that was blind. A group of Snake Indians had camped near the same hunting grounds; and when the Utah chieftain heard that his enemies were so near, he decided to go on the warpath. Calling on his big friend, he asked him to head one band of warriors. Smith graciously declined, not wishing to risk his neck, saying that he had no suitable horse for a war party as his was blind. The chief immediately had a beautiful horse completely equipped and brought to Peg-Leg, who now had no alternative but to join the war party.

Smith, at the head of one group of braves, approached the Snake encampment with the rest of the party. Utah war whoops broke out, and high above the rest was the clarion voice of Terry-oats-at-an-tuggy-bone. Smith's horse, being high strung, "cold jawed," and, despite all the efforts of his rider who tried to stop him or guide him to the right or left, kept galloping straight for the Snake encampment, bearing his unwilling burden. Over every obstacle the horse charged straight into the midst of the astonished Redskins.

Close in Smith's wake followed his native allies, trying their best not to be outdone by a paltry paleface. Smith, making virtue out of necessity, shot down the chief and grabbed a war club from his saddle, with which he dealt blow after blow on all sides. The Snakes, panic stricken by such a fierce and unparalleled onslaught, undoubtedly without peer in their historic annals, fled. From that day on, the name of Terry-oats-at-an-tuggy-bone was stamped as bravest of the braves.

In 1850 a traveler stopped at Smith's camp and found him the undisputed owner of hundreds of beautiful Spanish horses. Smith explained that he had obtained them with difficulty from the Spanish country (California), which had the best horses in the world. In fact, they were unusually expensive horses, he told the traveler. He had lost several braves and three of his Indian wives had lost brothers and one a father. There was no doubt, he asserted, that he had paid for all he drove away.

His raids seemed successful to the traveler, who asked him how many horses he had succeeded in obtaining. "Only about three thousand," he replied. "The rascals got back half what we started with."

Smith had a partner in this expedition who was almost as important and colorful as Smith himself, Chief Walkara, head of a band of Yampa Utes. McCall, who visited his tepee in 1840, describes Walkara as being big headed, short trunked, and bandy legged, but a superb figure on horseback, seeming as much a part of his horse as the centaurs of old. Since he was equipped with many fine horses and was used to every hardship, he terrorized the whole Southwest, although his primary stamping grounds were New Mexico, Utah, and Southern California. His movements were so rapid, his plans so ingenious and well executed, that there is no record of his having failed in any enterprise. His fighting costume was as novel and picturesque as his personality. When on the warpath, he would don a suit of finest broadcloth cut in the latest fashion, a cambric shirt, and a beaver hat. Over this he would wear the gaudy Indian costume. As he rode at the head of his braves with their colorful horses, the gaily embroidered saddle pads glittering with metal ornaments, he might have been taken for an Arabian chieftain of the Western deserts.

According to Antoine Robideaux, 6 white men and about 150 Indians took part in this grand expedition to the Spanish country. Jim Beckwourth preceded the company and made his headquarters at the Chino Rancho, pretending that he was going to remain to try his hand at sea-otter hunting. Using this ranch as his headquarters, he obtained all necessary information; and when Peg-Leg and Walkara appeared in the Cajon Pass, he was ready to counsel and assist them. Once the complete plans were made, the movements were carried out with the usual rapidity and success of the sly Walkara. Every ranch south from the Santa Ana River to San Juan Capistrano was plundered of its best horses and mares. One of the most spectacular of these visits was at San Luis Obispo, where the raiders succeeded in getting 1,000 tame mares and over 200 excellent horses, taking approximately 1,500 good animals from this mission alone. Needless to say, the wily chieftain was in charge of this particular raid. During the dead of night the band quietly cut an opening into the corral where the finest saddle horses were under guard and drove them triumphantly off into the night, leaving the guards astonished and the mission destitute of horses. In fact, it even had to send an Indian on foot for help.

California was in an uproar following this wholesale robbery. Officials and citizens hurriedly organized posses to catch and punish the offenders. So great was the public excitement that the prisoners were let out of the Los Angeles jail to form one party. However, the Californians' first encounter with Peg-Leg met with complete and shameful defeat, providing the trapper with a yarn for many a year.

Peg-Leg and his party found that, because of the great numbers of horses they were driving, their progress was so slow they would be overtaken. A plan was accordingly devised. The band split in two, the greater number pressing on with the stolen horses. The smaller band concealed themselves in the willows bordering a stream at which it was known the Californians would have to halt and quench their thirst and rest their weary mounts. As was expected, Palomares and his travel-worn men stopped for a brief drink and rest. As soon as they dismounted and threw themselves on their faces to drink, the concealed men quietly slipped out of the willows, mounted the pursuer's horses, and with a great clatter drove off all the rest of the horses belonging to the posse, leaving the group stranded on foot. The next day the men following Don Palomares picked up the stranded party and the pursuit was continued. It was rather easy to trail the herd of 3,000 horses and most difficult for the thieves to make any really fast advance. They were overtaken, and in the ensuing conflict the Californians succeeded in recapturing about 1,200 of the slower animals, the rest being safely taken away.

As a rule, stolen horses were sold in Utah or taken directly to Santa Fé. At that time little attention was paid to the brands of horses, and the animals were bought indiscriminately. Thus these horses of California were distributed throughout the United States both by the prevalence of horse thieving and by legitimate trade.

Tenacious "Red Head"

In 1848 the most astonishing of all the drives from the Pacific coast took place. This drive was under the direction of Miles Goodyear and was remarkable because of its climax.

Miles Goodyear had left home as a boy and was lured to the West by his adventurous spirit. Within a few months of the time he left home, he had become a member of a tribe of Utah Indians and had settled in the land made famous by Brigham Young. John

Adams Hussey, writing about Miles Goodyear almost one hundred years later, starts his account in this manner:

On the morning of April 23, 1848, a sentinel posted on top of a high hill near the Cajon Pass in Southern California, sighted several bands of horses approaching from the direction of Los Angeles. He immediately communicated the information to his commanding officer, Lieutenant George Stoneman, 1st United States Dragoons, who was charged with seeing that no stolen animals passed eastward through this important gateway from California into the Mojave Desert, from whence trails led to New Mexico and the frontiers of the United States. The Lieutenant, immediately sent forward a corporal with four men to halt the caravan at a narrow point in the pass. His report of a few days later laconically records what happened: "On the 23rd an American by the name of Goodyear arrived with 231 animals and 4 men. The animals I inspected and by my authority, gave him a passport. . . . I took from the American one mare not legally vented."

With these brief words the lieutenant recorded the data of one of the oddest of all the many drives out of California. It was not unusual for the lieutenant. In fact, he was stationed at the post for the express purpose of preventing any of the animals from being taken out that were not legally purchased. His reference to the animal "not legally vented" merely meant that the mare in question had not had her brand cancelled by a counterbrand when she was purchased. Each animal sold was always branded again by the owner with a special brand to show that he no longer had any claim to that animal. Each rancher possessed and recorded his own brand and his own vent.

Miles Goodyear was a frontiersman. He had built the first home and a trading post where Ogden, Utah, now stands. In 1847 he decided to sell out and go to California on a venture of his own. The Mexican War had broken out and the United States Army needed horses to mount the cavalry. In Mexican fighting, as in Indian fighting, the distances were too great for foot soldiers. Goodyear was quick to act when he heard that the quartermaster in Leavenworth was having difficulty in obtaining enough horses. He knew that horses were cheap and abundant in California, and he planned to take some from there to Leavenworth.

Early in the spring of 1848, Miles and his brother Andrew combed southern California for the best horses they could find. As later events proved, the Goodyears acquired horses full of endurance and stamina. They moved from the *pueblo* of Los Angeles to modern San Bernardino, through the Cajon Pass to the Mojave River approximately to Barstow, in the heart of one of the worst deserts in the United States. From Barstow they followed a route nearly the same as the modern Highway 91 to Las Vegas. The path in those days was known as *Jornada del Muerto,* and aptly so. Even today in a swiftly moving air-conditioned car, the journey is hot. When Goodyear reached Las Vegas, he pushed his weary horses to the Virgin River, then up the Santa Clara, through Mountain Meadows, across the Escalante Desert, and then north to the Sevier River. A branch trail led to the Great Salt Lake. Goodyear pushed on through the wastes of Nevada and Utah until he reached the main fork. Here he met the emigrants going west.

When Goodyear and his party finally arrived at Fort Leavenworth, they heard the news. The war had ended. The market had vanished. And they had driven horses more than halfway across the continent to this same market that now was glutted. Miles Goodyear then showed a spark of genius. He decided to turn the horses around and take them back to California. His reasoning was simple and obvious: Gold had been discovered since he left and horses must now be in demand in northern California. There was no market in Fort Leavenworth; there was one at Sutter's Fort, Sacramento, so there they would go.

Early the next spring the tenacious "Redhead" was on the trail. By this time, too, his unbroken horses were thoroughly trail wise and undoubtedly gave him a minimum amount of trouble. When first leaving California, they must have taken every opportunity to bolt for the thickets and any other likely-looking chance which might mean freedom. Now they settled down to the pace and shuffled along at a steady gait. In any case, their return trip from Independence, Missouri, to Sacramento, California, was made in fifty-four days—almost a record for a drive of that sort.

Miners in Sacramento said that the Goodyear animals were in a wretched state, and small wonder. Had they not just traveled barefooted over almost four thousand miles, most of it desert, with alkali water, and the last half at a forced speed? It was a superb test of horseflesh and horsemanship. It is of interest to note that in the end Miles Goodyear made a profit from the expedition. Goodyear settled in California, where he died and was buried at Benicia. Undoubtedly he felt he had seen enough country.

The drive of Goodyear may have been the most interesting, but it was by no means the last. Until the coming of railroad transportation, driving livestock east from the Pacific coast was a common occurrence. Daniel Bruhn was with some forty wranglers who drove three thousand horses to Denver in 1870. In the same decade Juan Sepulveda drove two large herds east for John Forster.

Since livestock was plentiful in California, it is not surprising to find horses driven out in large numbers. Some of the early drives had the Louisiana sugar-mill markets in mind. The Rocky Mountain men traded California livestock with the Indians for furs. Other animals were sold to the government, which needed cavalry horses. Part of the stock was utilized in the settlement of the Great Plains and Rocky Mountain states, as work animals, as saddle animals, as freight animals, and for the production of the mule for mountain transportation.

◄§ *Horseplay*

In Mexico a popular diversion was *correr el gallo*. This was a straightaway run, between two designated points, of a rider carrying a rooster. The rider was given a suitable start and then was pursued. Samuel Reid, Jr., tells of one of these races in which a Texas Ranger, who was patrolling the border, took the leading part:

Clinton Dewit, volunteered to bear the chicken to camp; and seizing the bird by the legs, dashed off at a break-neck pace. After he had got a fair start, about one-half of the Mexicans rushed after him, yelling like a legion of devils, the remainder ran by a shorter route to intercept him at different points on the road. We galloped out of town to see the sport. They had purposely chosen a road for him to follow that was covered with loose rocks, and full of holes, in hope that his horse would either stumble over the one or fall into the other. But Clint Dewit was too good a horseman to suffer either of these mishaps, and picked his course over the uneven ground as coolly as if he were gallanting a lady to church. The speed of his horse was so much greater than that of his pursuers that he was soon able to leave them far behind, but those who had cut across and got before him, annoyed him exceedingly, so that he was frequently obliged to run right over them, (which he always did when he had a chance,) or task his horse's powers to the utmost to ride around and avoid them. When he had nearly reached the camp, a big, burly fellow, mounted on a strong horse, rushed out from behind a house which he was obliged to pass and grasped the prize. So eager was he to secure the chicken that he momentarily released his hold on the bridle, while both horses were dashing along at full speed. "Clint" immediately perceived his advantage and grappling him by the throat, suddenly reined his horse up. The consequence was that the fellow's horse passed from under him, and left the rider in Clint's grasp. Dropping him to the ground, the young Texan clutched the prize, and raising a yell of triumph, bore it easily to camp.

In Texas there was another variation of the game in which a gander was hung to a tree by his feet. Then his neck was well greased with soap or tallow, and he was ready for the riders. The object was to pull off the head while riding at full speed. Money was often bet on the outcome, and the participants would sometimes each put up a certain amount of money to go to the first successful competitor.

Once in Texas during the early days two young men went out to hunt buffaloes. One of them, named Babb, discovered before long the riderless horse of his companion approaching him. He immediately began to search for his friend. Before long he came across him astride a huge buffalo bull. Babb went after the bull at full speed in order to help his friend off. When he drew alongside, he asked him what in the world he was doing sitting there. His friend's answer was that his horse could not run fast enough to catch the buffaloes, so he jumped on this bull that was going by in order to catch them. Babb asked if he wanted a hand to get off the buffalo, and the reply was, "Certainly not." Taking out his knife the rider stabbed the buffalo and, as he fell, jumped aside and remarked, "That's the way to get off an old buffalo bull."

Roping a buffalo was not only dangerous, it was almost impossible. The strength of the horse and man was not comparable to the strength of the buffalo. At Buffalo Gap, Don Drury and Hiram Crain of Brenham, Texas, came upon a two-year-old bull. Don wanted to rope it and in no time had the noose over the buffalo's head. He would no sooner stretch it out than it would bounce back on its feet. Finally, as the horse began to give out, Crain shot the bull. Don had to admit that he had taken in a little too much territory that time.

Probably the best-known example of the American cowboy's roping buffaloes occurred as a result of the Chisholm cattle trail, which came to Abilene, Kansas, from Texas. Joseph McCoy decided that the best way to advertise and gain a market was to ship a carload of buffaloes to the corn belt. Mark Withers, a trail driver out of Texas, said he would catch the buffaloes and put them in the stock car. It takes a lot to feaze a Texan. The resulting activity was nothing less than remarkable, even at that time when an occasional happy cowboy would not hesitate to drop his rope on the smokestack of an annoying railroad engine.

The Spanish Horse in South America

... el caballo criollo ... primero, fue un elemento de colonización, en segundo elemento de civilización, después el arma formidable de la ... guerra civil.—Dr. E. S. Zeballos, "*La Agricultura en ambas Americas* (1894).

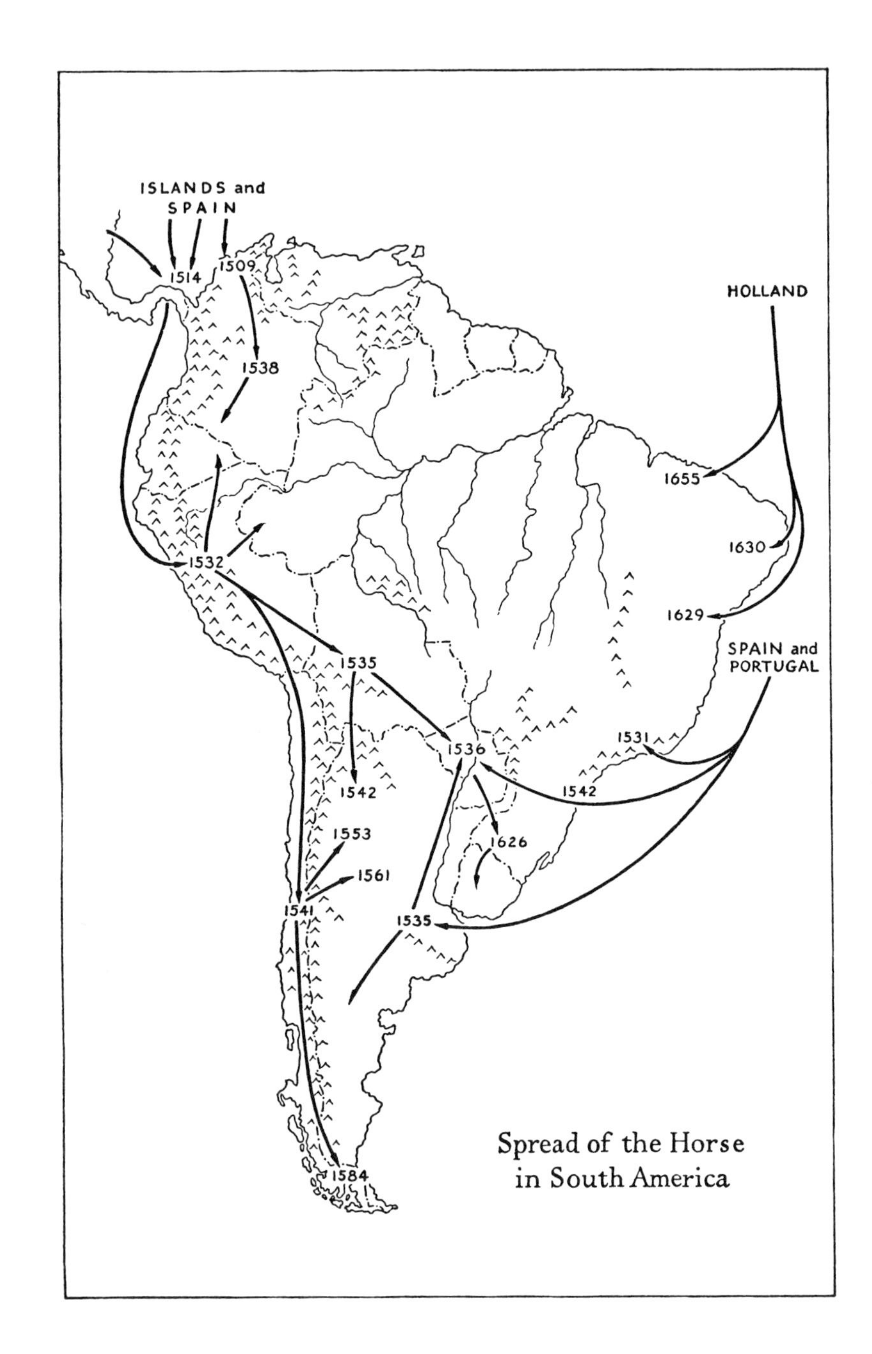

Spread of the Horse in South America

10. Our Swords in One Hand

Precocious Horses

Although all of the early chroniclers tell us something about horses, there are a few (apparently horse-lovers) who could never pass up an opportunity to tell of the part played by the horses in the conquest and development of the New World. Bernal Díaz del Castillo and Hernán Cortés are good examples in North America, while Garcilaso de la Vega and Agustín de Zárate belong to this class of writers on South America.

The Inca Garcilaso de la Vega, whose father was a *conquistador* and whose mother was an Inca princess, was born and reared in the New World and went to Spain only as a man. He never returned to his native land, finally settling down in Córdoba. With his retentive memory and his constant talks with returning *conquistadores,* his writings have an authentic ring that many of the early historians, writing entirely from secondhand sources and with a Peninsular outlook, entirely lack.

In Garcilaso de la Vega's *Comentarios Reales,* he speaks of the first horses to go to Peru. Characteristically, he states that the horses were taken to Peru by the Spaniards and that without their help the Spaniards could not have conquered the New World. Most of the horses spreading southward into Peru, he continues, were descendants of Andalusian animals transported to the islands of Santo Domingo [Hispaniola], Cuba, and Jamaica, and were taken from there to the Isthmus of Panama.

By 1526 horses were sufficiently plentiful in Panama that this area could furnish most of the beasts necessary for the conquest of Peru. Another colony from which Peru obtained a number of its first horses was Nicaragua, which since 1530 had been a dependency of Panama. Mexico also furnished some horses when Alvarado, persuaded not to join forces in conquest of Peru, returned to Mexico but left his horses to help in the conquest. Most of Pizarro's brood mares, on the other hand, came from Jamaica. There exists a capitulation between Pizarro and the Queen dated 1529 whereby the Queen gave Pizarro permission to obtain twenty-five mares and a few stallions from the royal ranches of Jamaica.

The horses which the island breeders had were good, and they bred more of the same kind. It had become obvious to many settlers that the need for horses in the immense stretches of this new land would be tremendous and the prices being paid for them were already good and promised to be better. Some ranchers soon had a number of brood mares on their establishments and were prepared to sell horses to the highest bidder.

Ricardo Cappa agrees with Garcilaso de la Vega regarding the origin and value of the early horses in Peru. Cappa, however, goes further and points out that, if the conquerors wished to import horses directly from Spain, it was first necessary to obtain permission from the Council of the Indies. Permission had to be obtained not only to buy them in Spain but also to transport them to the New World and to breed them afterwards. Governmental regulations are not a modern innovation. In fact, some of the early laws regulating horse affairs during Pizarro's time are rather interesting. The *Cabildo* of Lima fixed prices for many common transactions. For example, only ten pesos could be charged for burning a spavin; for judging a horse's age and health, eight pesos; for shoeing all four feet, one and one-half pesos; for bleeding, one peso; for trimming the mane and tail, two pesos.

According to the Inca, the horses of Peru were more precocious than those of Spain. He illustrates his point by asserting that when he rode with the lancers' cavalry in Peru (1560), he rode, as was not at all unusual, a two-year-old horse.

The Chilean historian and hippologist, Uldaricio Prado wrote one of the best accounts of the arrival and spread of the horse in South America in his book entitled *El Caballo Chileno.* According to him, by 1530 or 1540 horses were being raised in rather large quantities in the New World.

The horses which were imported directly from Spain must have been exceptionally good and very valuable. If they were not valuable, it would not have been worth the trouble and expense of bringing them on the long, dangerous trip, not to mention the difficulties encountered in obtaining the necessary cedulas and other permits and papers needed. The very fact that Pizarro did take the trouble to have some horses sent directly to Peru (such as those that Illan Suáres de Carbajal brought him) indicates the value placed on good horseflesh. These horses which he had imported were especially trained in jousting and the other equestrian sports which were popular at that time.

Like a Flight of Doves

Peru, with its mountains and valleys, proved to be an excellent area in which to raise saddle horses. The quality of the animals produced was soon noticed and commented on by the historians. Garcilaso de la Vega states that the way the horses would travel over the rough country was miraculous, adding that they moved with "the ease of a flight of doves." The Inca does not claim this comparison for his own but says that it was originally used by one of the conquerors he knew, Francisco Rodríguez de Villafuerte, who has been a companion of Pizarro.

The Inca also tells a story of an accident that occurred in 1583, when a large herd of horses had to be moved. One of the *conquistadores* who was living in Cuzco, Juan Julio de Ojeda by name, had a buckskin horse that was his favorite animal. On this particular day Juan Julio was riding his buckskin on the road to Villacori, acting as the leader for a large herd of horses, when his reins, because of a moment's carelessness, slipped to the ground.

The conquerors, like all horsemen who lead a dangerous life, rarely tied the ends of the reins together as is common today. When one gets off the horse rapidly (either on purpose or accidentally), it is much easier to hold him if the reins are split and fall to the ground with the rider.

As Juan Julio's horse was galloping rapidly over extremely rough ground, Juan Julio lost his balance and fell. The horse immediately stopped and so saved his master's life, for the herd of approximately three hundred horses were right on his heels. The animals apparently parted and ran to each side of the fallen man and his faithful horse. The historian adds, "And in this way was the man's life saved by the loyalty of his horse."

The qualities and characteristics of the early Peruvian horses are of special interest, for they formed the nucleus that was later to supply most of the horses of South America. The Peruvian horse, with the exception of a few direct imports, was the progenitor of the horse populations of Bolivia, Chile, Paraguay, and Argentina, as well as much of Brazil and Uruguay.

Peru, an administrative center, was the seat of the viceroy and thus the social and political center of the new land. A regular court, similar to the court of the king of Spain, surrounded the viceroy. The viceroy was the monarch in the New World, and there were continual ceremonies connected with his position. In these ceremonies and *fiestas* the horse played a very important role, and certain strains of horses were imported and bred especially for these functions. This practice tended to improve the blood, as great care was taken in the selection, breeding, and raising of the animals. It is also reflected somewhat today in the modern horse types found in Peru such as the *Costeña,* the *Serrana,* and the *Morochuca.*

There were a few direct importations of particularly valuable animals. Illan Suáres de Carbajal, for instance, brought horses especially for the ceremonies that were held when Francisco Pizarro obtained the rank of marquis. These importations tended to keep the quality of the horses high. As could be expected, some of the nobility who came to Peru from the Old World insisted on bringing a number of their favorite stallions and mares, as did Hurtado de Mendoza, who belonged to the highest nobility of Castile—his family was related to the Cid. Some of these nobles may have lacked formal schooling, but they had been thoroughly instructed since childhood in all subjects deemed necessary and fitting for a *caballero,* or horseman.

One such *caballero* was Don García, a relative of the viceroy. Apparently believing that home was all right for womenfolk but no place for grown men, he had left at seventeen to join the king's army in an expedition against Corsica. In this campaign he proved himself already a master horseman. A little while later he fought with Charles V during his disagreement with the French in 1554. Here, too, he proved himself a warrior and a horseman. Shortly afterward he accompanied his father, the Marquis de Caneta, to Peru. When he reached the ripe old age of twenty-two, he was appointed governor of Chile, and none of his men dared suggest that they were being led by a child. When Don García went to Chile, he took an exceptional group of horses and horsemen with him south to the land of the Arucanians. There were in this expedition 150 horsemen and some 500 horses.

During the early years of the conquest of Peru (that is, until about 1550) horses were so scarce that they were sold only upon the death of the owner or upon his return to Spain. During this time their value ranged from four to six thousand pesos. Beginning in 1542, the price began to drop as horses began to increase in numbers. Within two decades their value was down to three or four hundred pesos for a good horse *(buen caballo),* and indeed there are records of poor ones being sold for as little as twenty or thirty pesos.

One of the regions where the best Peruvian horses were raised in the early days was the province of Charcas, which bordered the southern part of the province of Cuzco and extended well south of Lake Titicaca, including much of present-day Bolivia. Roughly, its border extended from modern Paraguay westward to the Pacific coast at Copiapo. The rich Potosí mines were in this region, as also were fertile pastures and valleys such as Canela, which was part of the royal *encomienda* of Valdivia until 1540, when he decided to conquer Chile. In 1585 there were in the district of Jaen de Bracamoros four thousand brood mares on ten ranches. Before the seventeenth century Peru was able to raise the horses she needed and to start accumulating for future demands.

Charcas to Marga-Marga

The first horses were introduced into Chile in 1541 by the *conquistador* Don Pedro de Valdivia and were obtained in southern Peru. Uldaricio Prado says that the Chileans obtained many horses from Argentina, where they had been brought by the *Adelantado* Pedro de Mendoza. However, this is not very likely during the early days of settlement, although it did occur to some extent later. It is much more likely that Argentina obtained most of her original horses from Chile.

Don Pedro de Valdivia had in his original expedition seventy or seventy-five horsemen and about an equal number of foot soldiers. He also had pack horses, the majority of which may well have been mares as the men preferred to ride the stallions. Since there were not enough horses at that early date to make unlimited choice possible and mules were not readily available, the mares probably became the pack animals.

The gathering of these horses is reported to have been done

around Charcas. Since this province is known to have been one of the first settled and since it contained fertile valleys and stock ranches, the statement seems logical. Three years later Don Alonso de Monrroy also gathered seventy more horses in this region and took them to Chile.

Later, in 1546, eight Spaniards arrived in Serena with ten more brood mares also from Charcas. In 1547, García de Caceres and Diego de Maldonado brought sixty mares to Chile from the same Province of Charcas. In 1548, Valdivia imported eighty horses from Cuzco. In 1551, Francisco de Villagrán came to Chile from Charcas with four hundred horses. When one makes a rough summary of known importations, it becomes evident that there was a basis of around three hundred to four hundred mares available for breeding purposes at a relatively early date, and that most of these came from the province of Charcas in Peru.

By 1603 horse breeding was firmly established in the rich valleys where the extensive ranches of the colonizers of Chile were found. The occupation was quite profitable because of the continued need for horses by the army. When the military forces were not on an expedition, the animals were kept in training by the raiding Arucanian Indians. Horses were always in demand; and thus it can be seen that Charcas was the origin of most of the Chilean horses.

The first important Chilean breeder was the man who later became the first bishop of Santiago, Don Rodrigo González de Marmolejo. His ranch was called "Melipilla" and was in Marga-Marga. He also had holdings in the valleys of Quillota and Aconcagua.

With the arrival in Chile of the new governor, young Don García Hurtado de Mendoza, who was, as has already been explained, a horseman to the core, the continued existence of Chile and the eventual creation of a Chilean horse, was assured. Don García was equally capable of riding either in the old Spanish fashion of *a la brida* or in the more modern and popular *a la jineta* style, and he had outfits to go with each type of horsemanship, riding in whichever style was demanded by the *fiesta,* game, or joust. Since horses were his hobby, it is not surprising to find him bringing forty-two animals of his own. All of these were carefully selected—some for war, some for games, some for parade, and, more important, others for breeding purposes only.

☙ *Forced to Ride Mares*

The first civil regulations regarding horse breeding in Chile appeared in 1551. This early legislation clearly showed an appreciation of horses and the realization of the need for careful breeding. An official called an *yeguerizo* was appointed to assist the breeders and to see that the program was effective. There were even certain regulations which indicate that some sort of selective breeding was carried out by the government itself.

The scarcity of horses in Chile in the early days was not detrimental to their future development. The early soldiers and settlers were forced to ride both mares and stallions, despite their desire or tradition. Consequently, since both sexes were used equally by the owners, both had the opportunity of having their value proved by actual use, and only the best survived and increased. The incredibly arduous expeditions, the uncharted roads, the virgin and rough mountainous terrain, the turbulent rivers and deep gorges, all combined to make a tremendous demand upon the body and the stamina of the horse in Chile. A selection of individuals resulted. Frugality, probably after stamina the most valuable trait, became an outstanding characteristic. A contemporary writer who helped settle Chile said that when the Spaniards conquered and settled the city of Santiago, they lived for years in continuous warfare with the surrounding Indians. In order to survive, he said, "we had to keep our swords in one hand and our plows in the other." It would be difficult to find a passage which demonstrates more clearly the constant selection which went on in the early days. Another contemporary,

Alonso de Ovalla, wrote that the Chilean horse, in conformation, endurance, and other excellent qualities, was superior to the Andalusian. He stated that there was no reason for the horse to degenerate in such an admirable land.

After the first hazardous years of the conquest, Chile began to establish order and peace, and except for the fringes, where war was carried on with the Arucanian Indians, agriculture and ranching prospered.

The climate, the living conditions, and the geological formation of the many valleys in Chile, all acted on horse breeding in a very different way from many regions of South America. The horses did not have the freedom of the pampas, where they lived without the watch and care of man, as in Argentina. Neither did they suffer the difficulties that surround tropical life or life in the high altitude of the Andes. The small Chilean valleys permitted horse breeding to be carried out under the guidance and care of the rancher, whose personal desire and need for good horses may not have exceeded those of other South Americans, but certainly were more easily achieved.

The old Chilean *hacendado,* or rancher, in common with the other Spanish Americans, was a born horse enthusiast, and his special love was equine sports. He guarded his honor, his horses, his family, and his flocks in about the same order. All his pleasures and all his work was done on the back of the ever necessary horse.

The improvement of horses (principally by better selection of stallions, increased pains in training, and improved feed) occupied most of the time between 1610 and 1810. The tranquility of this colony, rarely disturbed by foreign affairs, resulted in a solid and enduring work, upon which the later selection for the national breed could be made.

The Chilean horse has been influenced somewhat by the Argentinian horse, although not so much as by the Peruvian. Argentinian horses were imported into Chile for the most part after the end of the seventeenth century (in fact, up to the present time). In the book *Orijenes del Ganado Argentino* by Don José A. Pillado, there are passages concerning the exportation of horses to Chile. In 1607 the Governor of Tucumán, Alonso de Riviera, complained to His Majesty of the difficulty in trading horses with Chile. This would indicate that there had been some trade between the countries. In 1609 Captain Pedro Martínez de Závala was authorized to buy 1,500 horses to be taken to Chile, and it may be presumed that he did get them. These two statements show that some horses were taken at an early date to Chile from Argentina. Two hundred years later a great many definite figures are available: for example, in the five years between 1860 and 1865, 7,607 horses were exported to Chile, and between 1879 and 1883, 6,267 were likewise disposed of.

11. Something Was Added

⸎ *Abandon the City*

Aside from the source material which is unavailable to most readers, there are few satisfactory accounts of the introduction of horses into the La Plata area in Argentina. One of the best readily accessible sources is the *Historia de la Ganadería Argentina* by Dr. Prudencio de la C. Mendoza.

The Argentine horse has a common origin with the horse of Chile, Paraguay, Uruguay, and Rio Grande do Sul, Brazil. Tucumán used to be a part of the province of Charcas, the area famous for horses in the early years of the conquest and settlement of Peru. As has been related, Charcas was a center of horse production and largely responsible for the original horses going to the rest of South America. Particularly is this true for Chile, Paraguay, and, of course, what is now Argentina. Diego de Rojas was sent south on an expedition by the Peruvian viceroy in 1542. On this journey he discovered Tucumán and was instrumental in introducing the first horses into that area.

The first horses in the La Plata region were those brought by the *Adelantado* Pedro de Mendoza. In a cedula dated August 12, 1534, and signed in Valencia, Spain, the King's secretary granted Mendoza permission to buy and load seventy-two horses and mares for his expedition to La Plata. They were embarked for the voyage at Cádiz in 1535.

The horses that Mendoza brought should have been good. He was a native of Guadex, in Granada, which was the home of the foundation mares of the world-famous Guzmán horses of Andalusia. He not only was a horseman but had grown up in blue-grass country.

After a long, difficult voyage, the expedition founding Buenos Aires landed on the coast of Riachuelo in 1536. There the Spaniards were graciously provided with food and provisions by the Querendi Indians who lived in the area around the mouth of the La Plata River. The friendliness of these Indians allowed them the opportunity of seeing and becoming acquainted with horses. Thus the Querendies lost their fear of the strange animals, which they had never seen before, much sooner than most Indians. They were not led to believe that the horse and the man were one animal as were some tribes in North America. The *Adelantado* Mendoza seems to have been too trusting in this respect. Probably he had not been made to realize the usefulness of the horse as a method of controlling the Indians, as had Cortés in Mexico and Pizarro in Peru.

Don Juan de Arjolas, a lieutenant of the *Adelantado,* was the first to take horses into Paraguay. He made an expedition up the Paraná River until he encountered two powerful Indian chiefs—Lambare and Nandu Guazurubicha—with 40,000 men. This meeting occurred just over the border of what is now Paraguay and may well have been the first introduction of horses into that country. Then, in 1537, a fort was established in Paraguay.

The fact that the Indians had lost all fear of the Spanish cavalry was only too apparent when the first hostilities broke out, when the Indians refused to feed the Spaniards any longer. In a surprisingly short time the Querendi Indians had become entirely familiar with the horse, and some were just as good horsemen as the Spaniards themselves.

The number of horses in the La Plata colony was considerably decreased a little later when the fort at Buenos Aires was sur-

Yellow Mount, a Paint Horse owned by Mr. and Mrs. Stanley H. Williamson, from a painting by Orren Mixer. (Courtesy Mrs. Stanley Williamson)

Two Palominos. Above: Charley Cole, a gelding foaled in 1967, owned by Mr. and Mrs. Richard McCarter. (Photograph by Margie Spence). Below: Mr. Pep 19, a golden Palomino champion, owned by the T Bar E Ranch, Argyle, Texas. (Photograph by Ray Bankston)

An Appaloosa, considered representative of the breed by the Appaloosa Horse Club, Inc. (From a painting by Marilee)

A Spanish Mustang filly of Bob Brislawn breeding, owned by Jeff Edwards. (Photograph by Jeff Edwards)

Bob Brislawn (center) and the author, photographed with a young *grullo* filly raised by Brislawn. (Photograph taken by Jeff Edwards at the National Convention of the Spanish Mustang Registry, Camino, California, 1973)

A *tobiano* Paint Horse. (Photograph by Margie Spence)

An *overo* Paint Horse. (Photograph by Margie Spence)

Leo Reno, registered number 1 in the International Buckskin Horse Association, Inc., registry, owned by Holiday Farm, Dyer, Indiana.

rounded by more than 20,000 natives, who had united for an attack. The Spaniards' need of food became so desperate during the siege that most of the horses had to be eaten. Some of their masters were also eaten before the Indians lost heart and withdrew.

In 1541 Mendoza decided to abandon the city of Buenos Aires. The Spaniards were evacuated in five vessels which had come for them from the settlement of Asunción in Paraguay. According to Rui Díaz de Guzmán in *Conquista del Río de la Plata,* when the cavalry was disbanded, the citizens were forced to leave five mares and seven stallions, which formed the basis for the wild horses that filled the plains of the Río de la Plata area. Although this is questionable, it is true that some thirty-nine years after the city on La Plata had been abandoned, General Juan de Garay refounded the city of Buenos Aires and found that wild horses had appeared in extraordinarily large numbers on the Argentine pampas.

Argentine Mustangs

Juan de Garay, during a voyage south of Buenos Aires as far as Patagonia in 1582 in search for the mysterious Trapalanda, found large numbers of wild horses. In a letter to the King dated in the same year, he also claimed that these horses came from those left by Don Pedro de Mendoza.

The horses abandoned by Mendoza at Buenos Aires in 1541, despite opinion to the contrary, seemingly were not responsible for the many wild horses encountered a few years later from Buenos Aires to Patagonia. In the first place, Mendoza had few horses left when he abandoned his colony; and in the second place, he used those he had for provisions. When one is looking for the real source of these wild horse herds, Peru or Chile furnish at least part of the answer. Santiago de Chile was founded by Valdivia in 1541. Immediately, a number of expeditions were organized and went out in all directions, both on missionary and on military pursuits. The greatest attraction was the much discussed but never encountered city of the Cesars which was the Fountain of Youth for Chilean explorations. To list just a few of these expeditions, almost any of which could have been responsible for some horses in Argentina, there was that of Diego Flores de León in 1541, of Francisco de Villagrán in 1553, of Pedro Sarmiento in 1584 (who got as far as the Straits of Magellan), and of Inigo López de Ayala in 1623. Also, the expedition of Argüello in 1540 cannot be entirely eliminated from the possibilities. One Brazilian writer claims the Portuguese settlement of São Vicente as the source of the wild herds—not an impossible, although unlikely, source.

Some Spanish and Portuguese writers have tried to prove that the wild horses of Argentina were not of European or Spanish origin but native to America. This claim is certainly untrue. The Argentine mustangs were a descendant of the Andalusian or Spanish horse. The predecessor of the horse in America never developed beyond the Pleistocene *Equus*.

The sudden appearance of the wild horses furnished a new table delicacy for the Querendies. It was not long until wild-horse flesh constituted their principal food. They showed a marked preference for the meat of the colts, which they probably liked as we like baby beef. Both the Indians and the Spaniards used the grease of the wild horses to make their campfires, the hides to make clothes, boots, and tents, and the manes and tails for decoration and ropes.

Characteristically, the Argentine mustang was of medium size, with large head, thick legs, fairly prominent ears, and a big neck. The coat generally was dark. Patched roan horses *(overos)* are mentioned, but pintos *(tobianos)* were apparently unknown or at least uncommon. The mustangs lived in great bands, often moving long distances during a single day when they were traveling. Sometimes they traveled single file in the manner of the Indians.

The range of the wild horse in Argentina was extensive, stretching from Tucumán and Corrientes to the Straits of Magellan.

☙ *Still a Pedestrian*

Paraguay (and a little later the Río de la Plata colony) received its second infusion of horses when the expedition of Álvar Núñez Cabeza de Vaca arrived after traveling overland across southern Brazil in the year 1542. Apparently the transcontinental hike Álvar Núñez made across the United States a few years earlier had not dampened his enthusiasm for cross-country travel. According to the stipulations of his contract, Álvar Núñez was to furnish horses to the Río de la Plata colony, and with his expedition there were forty-six horses and mares. He sailed from Sanlúcar de Barrameda on the second of November, 1540, and arrived on what is now the Santa Catarina coast in southern Brazil on the twenty-ninth of March, 1541. He landed and disembarked his men and the twenty-six horses and mares which were the only survivors of the long voyage. He left Santa Catarina on November 2, 1541. Crossing the rough country which lay between him and his destination, he at last arrived on the bank of the Paraná River on February 1, 1592. Assisted by the Indians, his party crossed the Paraná a little below the Iguazú Falls in canoes and barks. As he approached Asunción, he wrote to Domingo Martínez de Irala, who was then in charge of the colony. On the second of March he arrived at Asunción at nine in the morning, and there received all the honor due the high rank he held. The horses brought by Cabeza de Vaca must have been a welcome addition to those already in the settlement.

☙ *A Celebrated Quarrel*

By the time Buenos Aires was re-established, there were many wild horses on the Argentine pampas. These horses were a great inducement to would-be settlers, who well knew their value. The re-establishing of Buenos Aires in the heart of the dangerous Querendi Indian country was not easy, and only a great inducement could compensate for the many sacrifices which had to be made. The horses provided the stimulus which made the sacrifices seem unimportant. Thus the wild horses had an important influence in the re-establishment of Buenos Aires by General Juan de Garay. The General was grateful for the existence of horses which offered his men a return for the hardships they had to endure.

It is also of interest that until the time of the second founding of Buenos Aires there were apparently no cattle known in the La Plata area. From the date of the first founding until 1580, a period of forty-four years, there is no mention made of cattle. It would be presumptuous to think that the value of the pampas for general livestock-grazing had been overlooked. It was apparently just not feasible until the time of the second-founding.

A celebrated quarrel in colonial Argentina arose from a lawsuit between the last *adelantado* and the inhabitants of Buenos Aires over the wild horses found along the La Plata. It probably indicates more clearly than any other incident the importance of the exploitation of this resource during the early days of the colony.

In 1587, Juan Torres de Vera y Aragon, the crafty *adelantado,* arrived at his governmental seat in Asunción. There he was informed that Garay had founded in his name, somewhat to the south, the city of Buenos Aires. He also heard that there were large numbers of wild horses near Buenos Aires. Those who had seen them said there were at least 80,000 head. He was also informed that Garay had given the colonizers the right of capturing the horses for profit. It is hard today to believe that the *Adelantado* did not also know that Don Juan de Garay had to offer his companions the right of exploitation of this resource in order to repay them for the sacrifices imposed by the refounding of that city. Nevertheless, the *Adelantado* Juan Torres decided that the money brought in by these horses should be his; therefore, he claimed that they belonged to the royal ranches and were part of the royal property. He ordered that all caught should be sold at public auction in order, as he put it, to increase the King's benefit.

The settlers in Buenos Aires, with what must be admitted was good reason, considered this an usurpation of their just rights by the greedy old *Adelantado.* They said that they were the lawful owners of these animals because the right had been promised them by General Garay before they had even considered refounding the city. They mentioned the great amount of work done by them in order to rope and catch the colts and stated that the wild horses provided the only means of livelihood for the colony.

The people of Buenos Aires retained two of the best attorneys in South America—Pedro Sánchez de Luque and Gaspar de Quevedo—and put their case directly before the *Audiencia* (or court) of Charcas, for Argentina was under the jurisdiction of the viceroyalty of Peru at this time.

The *Adelantado,* on his part, told his lawyer, Francisco Pérez de Larinago, to press his claim. Long interrogations and judical investigations served as a base for the judgment eventually rendered by the tribunal at Charcas. The first decision was given on August 12, 1587, and the second on September 30, 1591. These decisions ordered the *Adelantado* Juan Torres not to take the wild horses that the citizens had captured away from them nor to hinder in any way the pursuit of more mustangs.

The *Adelantado* appealed the judgment, but without success. All of the above trial can still be read today, as the documents have been preserved in the *Archivo Generál de la Nación* in Argentina. In this same respect, it is of interest that on October 16, 1589, during the time of the quarrel and litigation, the *Cabildo* of Buenos Aires recognized the rights of the heirs of the original settlers to the wild horses. There is also preserved a royal provision of September 30, 1591, ordering the return to the citizens of Buenos Aires of all horses that had been taken from them by the *Adelantado.*

In 1591, as an outgrowth of the dispute between the citizens of Buenos Aires and Juan Torres, the *Adelantado* resigned his high post. And after receiving the favorable decision, the citizens of

Buenos Aires devoted all their free time to the exploitation of the wild horses, until at last the *Cabildo* of Buenos Aires found it necessary to take certain steps to limit abuses, especially needless slaughter of animals.

Tucumán

Following a directive issued by the Viceroy of Peru, one of his lieutenants, Diego de Rojas, discovered Tucumán in 1542. As Rojas was killed by the natives, further conquest of this part of northern Argentina was delayed. General Juan Núñez de Prado also came from Peru and founded in 1550 the city of Barco in the same province. Later the people of Barco were moved to Santiago de Estero, which thus became the oldest settlement in Argentina. General Prado had everything with him necessary for stock farming. In the National Library in Buenos Aires there are copies of documents regarding the founding of the city of Barco. One of Prado's lieutenants, Juan de Santa Cruz, brought more horses to Barco. The *Licenciado* Palo y Villagrán also went to Barco with men and horses. According to *Historia de Ganadería Argentina* by Prudencio de La C. Mendoza, Tucumán provided an ideal spot for agriculture, and as the Indians were relatively friendly, this colony thrived and prospered.

In spite of Rojas' earlier visit, it was probably Prado's expedition which was responsible for bringing sufficient horses to commence breeding. Diego de Almagro also brought some horses during his expedition to Salta. Some historians claim that in one battle he lost all of his horses, which could mean that some escaped.

Chile also was interested in the Tucumán area. Valdivia, in December of 1553, was appointed lieutenant of the city of Barco, which lies just over the Andes from Santiago. However, Francisco de Aguirre was largely responsible for establishing livestock in much of the zone of northern Argentina and Chile. The *conquistadores* of Chile introduced horses into Mendoza, which was founded in 1561. Northwestern Argentina was settled, about the same time, by people from Peru, Chile, and Argentina. Unlike many early settlements, it started on an agricultural basis and remained agricultural and prosperous.

The Argentine horse, after a long evolution and adaptation to Argentine soil, gradually developed into the modern *Criollo* horse. The climate, the pastures, and all the favorable natural conditions in the province of Buenos Aires and in the pampas helped develop this new American horse breed.

The Querendies, when they learned to ride, appreciated the ability the horse gave them in traveling over the open plains. The horse helped conquer the Indian and then gave him an appreciated helper for all the activities on the *estancias* of Argentina.

Dr. Zeballos in his historical study on the Agriculture of the Americas states that the *Caballo Criollo* was the first essential element in the civilization. He should also have added in the exploitation of the native Argentine people.

O Rio Grande do Sul

The horses of Rio Grande do Sul (Brazil) came jointly from Paraguay, from the La Plata area, and from the northern settlements of Brazil.

The honor of having brought the most, if not the first, livestock into Rio Grande do Sul, undoubtedly belongs to the Jesuit missionaries who had established themselves in Paraguay in 1586. From Paraguay, the fathers next settled what is now southern Rio Grande do Sul and northern Uruguay. Their missions flourished there during the seventeenth century, and their watchful care was responsible for the spread of livestock throughout the region, from the highland pastures of Vacaria to the level plains of Uruguaiana.

When the Portuguese settlers along the southern coast of Brazil were looking for an overland route to the colony of Sacramento, they encountered in Rio Grande do Sul large numbers of semiferal horses and cattle. These animals undoubtedly arose at least in large part from the livestock brought to the area by the Jesuits. They were utilized by the colonizers of the Gaucho state to stock their ranches at the beginning of the eighteenth century.

Unquestionably some livestock reached Rio Grande do Sul from the La Plata region. All of the early accounts speak of the numerous herds of wild horses in the La Plata area. Except for a few rivers, which were impassable only during certain seasons of the year, there was little else to hinder their following the extension of the pampas north and east from Argentina into Uruguay and Rio Grande do Sul.

The third important source of the livestock in Rio Grande do Sul came from the Portuguese settlements. Martim Affonso de Souza arrived in São Vicente in 1532 with an expedition well provided with animals. The Portuguese, like the Spaniards, could see the value of livestock to their colonists. The livestock of Portugal was for the most part similar to that of Spain. Since Portugal had also felt the influence of the Moorish invasion, there is little reason to believe that Portugal's livestock should be essentially different from Spain's in the sixteenth century.

One would think that Brazil was in itself large enough to hold these original Portuguese importations, but the Argentine writer, Ricardo Pillado, suggests that the original Portuguese stock formed the basis for the wild herds in Argentina. It is entirely possible that they helped stock the Argentine pampas, but it is more likely, however, that the livestock went both ways. Oliveira Vianna, another writer, pursuing this same theme, says definitely that the colony of São Vicente was responsible for the spread of livestock into Uruguay and Argentina. He traces the route as going across the mountains to Asunción and from there to the level plains of southern Paraguay and Argentina. It would then follow, according to his thesis, that the horses and cattle brought to Rio Grande do Sul from

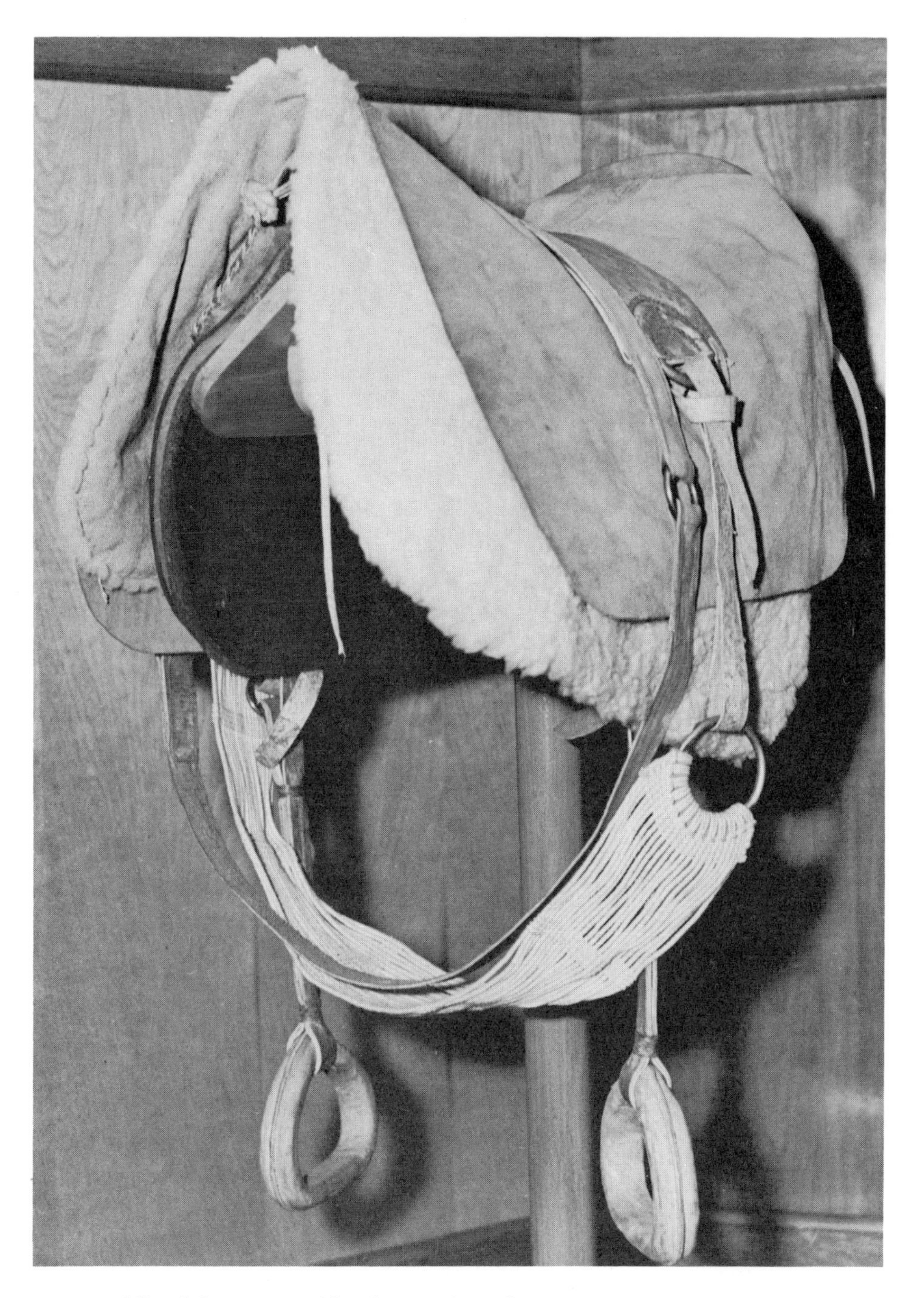

Saddle of the type used by the gauchos of Argentina, Uruguay, southern Brazil, and Paraguay. The saddle is covered with sheepskin. The riata is tied to a ring at the side of the saddle. (Photograph by Lucille Stewart, Los Angeles)

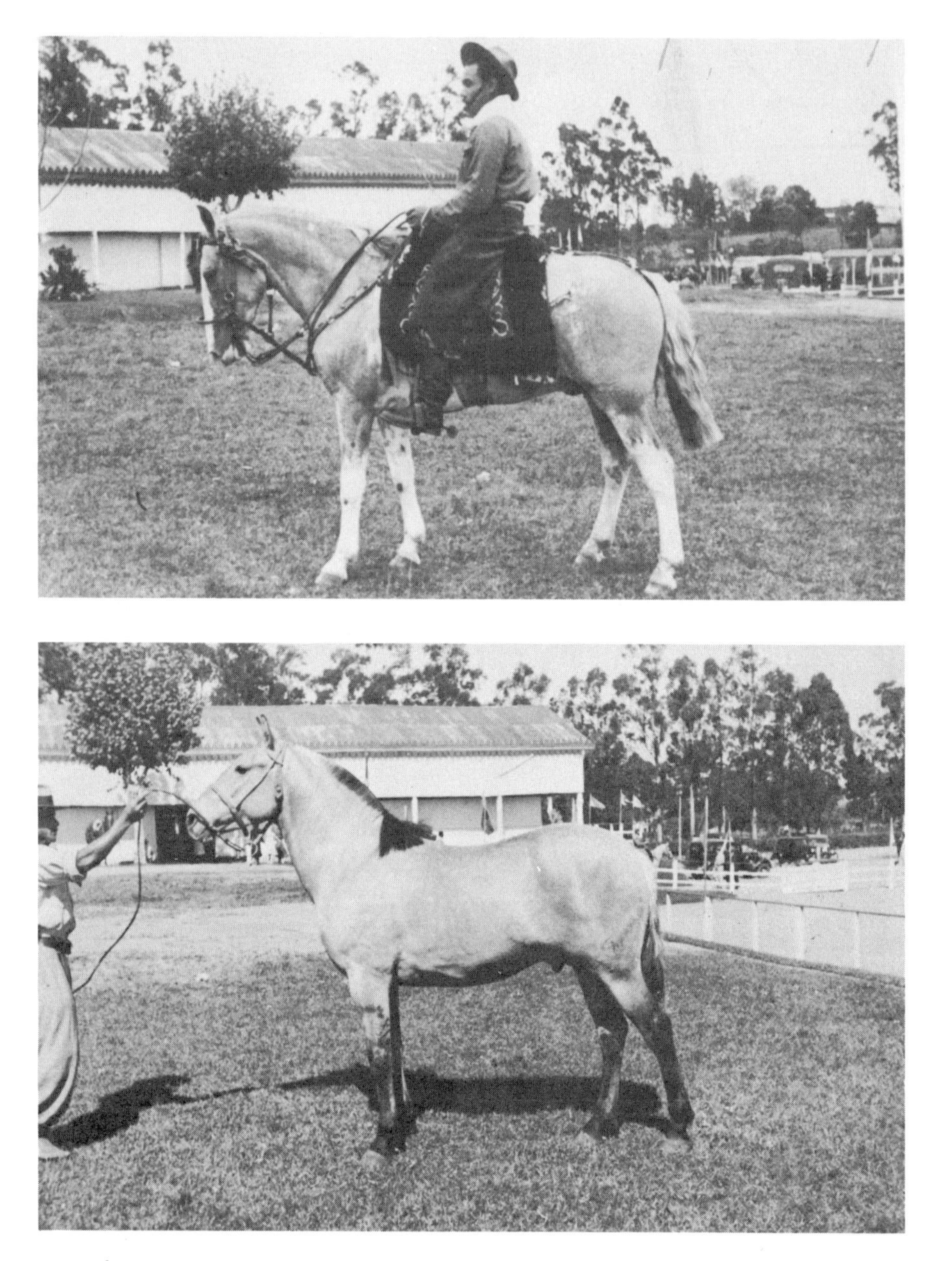

Champion *Crioulos*, Bagé, Brazil. (From photographs in the author's possession)

Paraguay by the Jesuits were of Portuguese origin from the first. This theory does not seem too plausible.

The *Crioulo* of Rio Grande do Sul may safely be said to have the blood of both Spanish and Portuguese horses. Certainly when we get back far enough, we will find Barb and Arab blood influencing the Brazilian *Crioulo,* just as the Moorish invasion influenced all Portuguese economy.

Added Color

The Dutch settlements had a great influence on the horse in Northern Brazil. The first Dutch or Flemish post in Brazil was established at Salvador, Bahía, in May, 1624, and lasted until April 30, 1635. A second establishment was founded on February 14, 1630, at Olinda. During this period a long stretch of the coast was occupied and the conquest of Pernambuco started. The Dutch controlled, by 1654, the land from Parahyba in the north to Bahía in the south.

In these settlements, especially the second, horses were introduced from the Low Countries of Europe and from southwestern Germany by the Dutch settlers. The Prince of Nassau, one of the governors of this colony, was known as an expert horseman. Records are available of activities and games on horseback carried on by the Flemish colonists and their Portuguese neighbors.

One of the main reasons why many native or *crioulo* horses in South America are multicolored is that Friesland horses were imported from Europe by the Dutch of northern Brazil. Before the end of the nineteenth century, there are very few, if any, references to pintos *(tobianos)* in Uruguayan and Argentine history. Thus there is a basis for giving some credit to the story that the *tobiano* pinto came to Uruguay and Argentina by way of Brazil.

The word *tobiano,* according to this account, is derived from an individual by the name of Tobias, a São Paulo revolutionary who was forced to flee south to Rio Grande do Sul in 1842 when he was defeated in the north. His arrival there with his men, some on pinto horses, gave birth to the term *tobiano*. Although, like many other stories, this cannot be proved, sufficient authorities and old inhabitants corroborate the statement to make it seem plausible. *Tobiano* refers to a clean-cut pinto and not to a paint, as many Westerners call the patched-roan horse, known in South America as an *overo*. In the United States these two types are often confused, and a better color terminology should be utilized. Even the English terms "piebald" and "skewbald" are not proper as translations.

The American Horse Today

He crossed every divide, rode into every coulee,
swam every stream.

12. Breeds, Strains, Types, and Colors

☙ *Heyday of the Cow Horse, the Mustang*

The "Great American Desert" was relegated to the Indians by Monroe, was considered in the days of Jackson as a fortunate barrier prohibiting Americans from straggling across the country, and was repeatedly characterized by men such as Catlin as being then and forever useless to cultivating man. Then, suddenly, undaunted, into the Great Plains came the mustang as a cow horse, leading and flanking vast herds of cattle, viciously snapping at any animal which dared stray from line. From the immense stretches of Texas, the half-broken mustang pushed the cattle northward into the land of the Indian nations and toward a railroad which was pushing into the waste. Farther and farther the cattle went, searching for markets and pastures, until finally they had spread from Matagorda to Manitoba. It was this natural-born cow horse which ended the reign of the Great American Desert, making that land his own. In search of grass he crossed every divide, rode into every coulee, swam every stream. A land was found which could breed wealth; cattle companies sprang up like mushrooms after a rain until no vocation could compare with the calling of the cow horse in its halcyon days during the last third of the nineteenth century.

Settlers, cheap barbed wire, the difficulty of segregating "customary ranges" on crowded land, and other factors finally led to fencing. Fewer cow horses were needed in fenced pastures, and improved breeding was possible. In place of the ubiquitous "longhorn," a white-faced Hereford appeared. Although the cattle ranch of the old days and its picturesque life were gone, the cowboys and the cow horses were still present and the cattle baron was still the nabob of the American uplands. Stock horses were just as necessary in the handling of cattle as ever before.

The drive of the Texas men has become a saga. Picture gaunt, dust-caked, nervous longhorns coming over a rise in the prairie with dilated nostrils eagerly twitching for the smell of water—a long black line disappearing into a trail of dust. Will James caught some of the elemental restlessness of this scene in his *Trail Herd*. Nor did he forget to put into the picture the ever present, all important cow horse. The long drives carried the cow horse of the Southwest to wherever range cattle are found. Four hundred years of heritage had at last blossomed. The heyday of the cow horse occurred during the decades between 1866 and 1885. Without him and the longhorn, part of America could never have been built. The latter part of this period saw this same cow horse changed in some localities almost beyond recognition. His blood was being diluted by driving, draft, and running horses brought in by Easterners, a people to whom "appearance" and "value" were almost synonymous. It was too much to expect a people used to tall, slender running horses or heavy draft horses to believe that an animal not much larger than a stocky pony could have the intelligence, stamina, and endurance to outwork on the cattle range half a dozen of their animals, to say nothing of one little feature called "cow sense," which the Eastern horses coming from northern Europe seemed to lack.

In the mountain ranges of the Northwest the new cattle owners thought that there was a demand for large horses to carry weight in the rough country, but with each gain they made in weight, there was a corresponding drop in endurance and cow sense. In Oregon, Clydesdale sires were used, in a few cases it must be admitted with fair success, but they were commonly called the "Ore-

A group of good cow-horse mares. (Courtesy *The Cattleman*)

gon Lummox," a term no doubt most apt for the majority. In Montana and the Dakotas, Percherons were used with identical results and alongside of the agile cow horse doubtless deserved their appelation, "Percheron Puddin Foots."

In California, led by the famous Miller-Lux outfit, which boasted they could drive from Canada to Mexico and camp every night on their own land, the mustang was crossed with Morgan blood and again a larger horse was produced. Many of these animals became good cow horses. Dutch economy had found a way to have a plow horse when age had slowed down agility.

Most polo ponies were bred by using "Thoroughbred" stallions and cow-horse dams, the result being as good a polo horse as the breeders are likely to produce. But then, as Cunninghame Graham says, "Polo players ride for pleasure and their horses are fed and pampered like Christians."

By crossing the Thoroughbred, the Morgan, the Clydesdale, and the Percheron with the lowly cow horse, breeders produced a taller and heavier horse, one that could run faster, or one more fitted for the parade or for the plow. Nevertheless, as it may seem, these are not the only qualities for which horses are bred. The cross was larger, faster, or stronger but soft and unfit to work with cattle. When the cowman had to travel over rough country with perhaps no food or water all day, and then, work over, follow a bunch of steers home under the stars, or sleep in the open on an icy night, or with a scorching wind drying out the grass, it began to dawn on him that his new half-breed was not equal to the work. Besides being clumsy on the turn, slow off the mark, and unsafe for running in rough country, he was an inferior weight-carrier. A cold night or two, without his blanket and his grain, made a poor horse of him.

While the breeding in the northern rim of the cow country was for weight, in the Southwest a new factor entered the picture, a factor which was to improve on the old Spanish cow horse—the

Roping ranch horses, an early photograph from West Texas. (Courtesy *The Cattleman*)

Quarter Horse. The "short horse" had features which the rancher of the West valued in saddle animals. He could both run and work. The Quarter Horse had been selectively bred for generations to produce an animal which would fairly fly over short distances. The mustang, on the other hand, had been bred by the merciless law of the survival of the fittest until his endurance and stamina were probably unequaled. Both could trace their ancestors to the Spanish Arab and Barb blood. The cross of the Spanish cow horse and the Quarter Horse improved both animals. The cow horse got a bit more weight, a new burst of speed, and some of his hard angles smoothed out. Now a horse had arrived that was considered by many cattlemen the greatest cow horse ever developed. Speed to overtake the fastest calf, weight to hold the heaviest steer, endurance to work day after day, and finally a desire and love to work with cattle were all combined in one animal.

Colonial Aristocrats

Little is known of the history of the horse in the original thirteen colonies, but a few facts are available. Wood, Sandys, and Gookin were the first to seriously import English horses into Virginia about 1620. Soon thereafter Governor Nicholson legalized horse racing, which immediately became popular, and by 1690 large purses were being offered. For several reasons, among them the lack of race tracks and straight stretches of road, it became the habit to run short races, generally along the main street of a town, as it was the only straight and cleared stretch available. J. F. D. Smyth, who was making a tour of the colonies right after the Revolutionary War, said, "They are much attached to quarter racing, which is always a match between two horses to run a quarter of a mile, straight out . . . and they have a breed which performs it with astounding velocity."

Short racing was also popular in Rhode Island, where some of the best running horses in the colonies were raised. William Robinson, one-time deputy governor of Rhode Island, bred some of the best running horses in that colony. His original sire, Old Snipe, according to an unsigned manuscript, was found in a drove of wild horses on Point Judith. Although Robinson did not realize it, Old Snipe's ancestors were Barbs, probably bred in Andalusia or Córdoba in Spain. It was not long until Robinson horses were famous for their speed. When an intercolonial match was arranged between the horses of Virginia and Rhode Island, so successful were the latter horses that the Virginians obtained some of Old Snipe's progeny to improve their horses.

The Virginia horses already had some infusion of Spanish blood, as there were Spanish horses in the backwoods of the colony, some wild and some owned by the Indians. This fact is borne out by a description of the Virginia horse at this time, a traveler from England describing them as "not very tall but hardy, strong and fleet."

The Spanish horses did not come from De Soto's expeditions, as had been commonly supposed, but from Spanish Guale. This district, established by Pedro Menéndez de Avilés, composed of the southeastern section of the present United States, had by 1650 seventy-nine missions, eight large towns, and two royal ranches. The predatory attacks of the English and the Indians on the Spanish settlements spread the livestock far into the north. Soon most of the southern English colonies had wild horses in the backwoods, and capturing them became a favorite sport among the youthful colonials, especially in Virginia and the Carolinas. Wild cattle were also captured in "cow pens" by the colonial "cowboys"—already so called in the eighteenth century—riding their fast-running quarter horses.

The Spanish horse moved north to the American Colonies with the assistance of the two main Indian tribes that bordered the Spanish settlements, the Cherokees and the Chickasaws. Both of these tribes were ideally suited for raising horses. They were inclined to permanent settlements, pursued agriculture to a certain extent, and had developed advanced societies. It was not long until they gained fame for their horse-breeding operations. A horse originating in one of these tribes could demand a premium when sold into the English Colonies.

Dr. John B. Irving, when speaking of the Southern Colonies, said that before the year 1754 the horses most highly regarded as saddle animals were the Chickasaw horses, which had originally been introduced by the Spaniards. He described them as small, around thirteen to fourteen hands in height, but remarkable for their muscular development. When crossed with horses imported from England, he added, they produced offspring of great beauty and speed.

J. F. D. Smyth, who traveled extensively through the Southern Colonies after the American Revolution (1784), wrote in his book that the Chickasaw Indians "have a beautiful breed of horses amongst them, which they carefully preserve unmixed."

Still another early reference to the excellence of the Chickasaw horse is found in the *South Carolina Gazette* in the issue of December 3, 1763. This quote has to do with a horse race won by a Chickasaw mare:

> Last Thursday, a great sweepstakes was run for, round the course at Edmonsburg, Ashepoo, the best of three heats, by horses, etc., which came in as follows:
>
> Mr. Edmund Bellinger Jr.'s Chicasah mare Bonnie Jane, rising 5 year old. 1, 2, 1.

Mr. Gibbs' Buzzard (quarter part English blood)	2, 1, 2.
Capt. Smith's English horse	dis.
Mr. Cochran's Chicasah h. Childers	d.
Mr. Coachman's grey gelding	d.

The last heat afforded excellent sport, as the jockeys whipped from start to ending post.

Since these Chickasaw horses are found in the background of most Western breeds, it is necessary to know something of the horses brought to America by the Spaniards. It is generally conceded that they were of Oriental origin. Spanish horses at that time were commonly referred to as Barbs or Arabs. The Moslem invaders who conquered Spain in the eighth century came from the Barbary States of North Africa. Contemporary writers refer to the horses as originating from African, Moorish, Barbary, or Tlemçen stock. They were seldom referred to as Arabs. The Arab horse of the eastern Mediterranean had a more refined, dish-faced head and a shorter back than his descendants in North Africa. The Spanish horses that came to America in the early days can be grouped under the general term Barb. Their blood, when combined with the English imports, helped create the quarter running horse.

The Quarter Horse

The Colonial short-distance horse was established in the American colonies at a date too early to allow the English Thoroughbreds to influence his breeding, and later, when he was raced against them, he would beat them on short distances because of his marvelous start. It must have been disconcerting to the Thoroughbred owners to import their running horses, those greyhounds of English turfs, and have them left at the post by the thick-set, close-coupled animals of the colonies. Janus was one of the few horses imported at this time whose blood influenced the quarter-of-a-mile running horse. He was imported from England by Mordecai Booth and landed in Virginia in 1752. According to the early records, Janus begat horses more noted for their speed than for their bottom. He was remarkable for his power and strength and compactness of form, and for his ability to transmit these characteristics. William Anson, a Texan and one of the earliest stockmen to interest himself in the history of the Quarter Horse, says, "From all accounts and the number of Januses which appear in all pedigrees, he must have been as prolific as our own Texas Steeldust, undoubtedly the most prolific horse which ever stood on four legs."

Racing in the Anglo-American Colonial period, as has been explained, was not a long-distance affair. The very absence of circular tracks and the scarcity of any cleared stretches made the long races inadvisable both for the jockeys and the spectators. American distance racing did not become general until the nineteenth century, and not until 1850 did the four-mile heats, which became very popular, reach their peak. Short races and "short horses" were the Colonial specialty. For them the short, thick, stocky quarter running horse was developed. Ability to start was much more important than ability to stay. The "short horse," or quarter-of-a-mile running horse, was developed at the same time as the American Thoroughbred. The first American Studbook of the Thoroughbred horse openly included "short horses" as foundation animals. Regular breeding of "short horses," or Quarter Horses, was discontinued by many Eastern breeders at the beginning of the nineteenth century in favor of a horse which could run longer distances. The horse then developed was the present American Thoroughbred, the greatest distance horse the world has ever known.

The Quarter Horse undoubtedly had much to do with the founding of the American Thoroughbred. One of the facts which demonstrates this is that many horses and mares in the first two volumes of Bruce's Studbook are described as Quarter Horses in Edgar's Studbook.

Ariel, the type admired by western horse breeders. (Courtesy Mereworth Farm and J. A. Estes of *Blood Horse*)

It is even possible that Justin Morgan was a Quarter Horse. William Anson said that this was generally acknowledged, adding that Stillman and members of the Morgan Horse Club of New York admitted that he could have been nothing else. Colonel Grove Cullum, an authority on Western horses who was once head horse buyer for the Remount Service and then a member of the New Mexico Racing Commission, said that, inasmuch as quarter racing was popular in Colonial times, it was possible that Justin Morgan was a Quarter Horse.

Although the short horse as such lost his general popularity in the East at the beginning of the nineteenth century, there were breeders, particularly on the Western frontiers, who continued to raise Quarter Horses. All through the years, beginning with Janus in 1752, Thoroughbreds of the right description have been crossed with the short horse.

The reasons for this fairly constant infusion of Thoroughbred blood are several. The Quarter Horse has a tendency to become too "short" in running ability, and a certain amount of coarseness sometimes appears. Thoroughbred blood lengthens his distance and gives him more refinement. Thoroughbred blood also helps to counteract a slight but common tendency toward mutton withers and stubby pasterns.

The most important stallion to influence the Quarter Horse after Janus was that great son of Diomed, Sir Archy. In many ways his importance is surprising, as he was a large horse, from the short-horse man's point of view, since he stood sixteen hands, and his best times were made on the four-mile tracks. Yet his blood, when crossed with a Quarter mare, left little to be desired. Many of the greatest short horses of the nineteenth century, such as Cold Deck, Shiloh, and Billy, trace back to Sir Archy. Even two of the best of the twentieth century sires, Joe Bailey and Peter McCue, are descended from Sir Archy. Joe Bailey traces three times on his sire's side and three times on his dam's side to this great sire.

A statue of Justin Morgan at the United States Morgan Horse Farm, Middlebury, Vermont. (Courtesy *The Cattleman*)

Some of the best modern short horses also have closeup Thoroughbred blood. In the 1940's and 1950's a large percentage of these were descendants of Chicaro, the Thoroughbred sire of Flying Bob; Port Drapeau, the Thoroughbred sire of My Texas Dandy; and Joe Blair, the Thoroughbred sire of Joe Reed. Today one finds the blood of Three Bars, Top Deck, and Depth Charge in a large number of the fastest quarter running horses.

The Quarter Horse (at least until recently) has been primarily a utility horse—a horse that has been owner-raised, owner-trained, and owner-raced. He has been the common man's horse. His utility has not been confined to the track, probably the reason for some of his weaknesses as a running horse. For three hundred years he has herded cattle, pulled stumps, and planted cotton six days a week—and on the seventh he has raced.

The Mustang Today

No one, for obvious reasons, was ever able to count, or even estimate accurately, the numbers of wild horses that once roamed the Western plains. One thing we do know is that their range extended farther than that of the millions of buffaloes. They grazed from the high mesas in Arizona to the verdant pastures of Montana. They were common from the badlands of the Dakotas to, as J. Frank Dobie said, "the ground across a thousand miles of Texas and Oklahoma ranges into eastern swamps that the buffalo did not sniff." Since no one could number them, they remain numberless. Today it is a different story: they are few and far between.

What happened was that the range of the mustang was gradually pre-empted by farmers and cattlemen. Any that hung around were rounded up, sold, or shot. The influence of the wild horse lingered on a little longer than that of the buffalo. The wild horse

A Quarter mare and foal. (Courtesy *The Cattleman*)

Above: Little Joe, Jr., an example of a good Quarter Horse. Below: Squaw H, a champion Quarter running mare.

could cross with domesticated horses, and their progeny were fertile. Consequently, mustang blood is found in most Western horse breeds to a greater or lesser degree. It is anyone's guess how many pure-blooded mustangs were still in existence by the early 1900's. Certainly there were not many. Those that survive undoubtedly contain blood from animals that escaped roundups or that were lost or deliberately turned loose. Since they were not a game animal and were often a nuisance to agricultural interests, they were shot, driven away, and even rounded up and sold as dog food.

Their present semiprotected condition is due in a large part to the campaign waged almost single-handedly by a woman, often called "Wild Horse" Annie. Her real name is Mrs. Velma B. Johnson. From her home in Reno, Nevada, she took her fight to the halls of Congress and won. In a letter dated March 5, 1968, she explains how her role differed from that of the mustang breed associations:

> I do not consider myself an expert in that field [the Spanish mustang]. I have always concerned myself with *all* of the wild horses. My knowledge has been confined to the over-all picture of the harassed, abused, remnants of our Western heritage.

Mrs. Johnson's battle with the United States Bureau of Land Management and the Department of Agriculture (Forest Service) was long and hard, but, owing mainly to her efforts, the plight of the surviving wild horses looks better than it has since the 1860's.

J. Frank Dobie wrote, in his interesting book *Tales of the Mustang* (published in 1936), "There are a few wild horses left, but no pure blooded Spanish Mustangs, left on the ranges of North America." Perhaps "Pancho" Dobie was wrong this time. At least, there are (and were) some old-time mustangers, like Monty Holbrook, Ilo Belsky, Bob and Ferdinand Brislawn, Gilbert Jones, and Charles and Dick Williams, who would have disagreed. They were

Mustangs, a sculpture by Alexander Phimister Proctor, for the University of Texas, 1948. The words engraved on the base of the statue: "These horses bore Spanish explorers across two continents. They brought to the Plains Indians the age of horse culture. Texas cowboys rode them to extend the ranching occupation clear to the plains of Alberta. Spanish horse, Texas cow pony and mustang were all one in those times when, as sayings went, a man was no better than his horse and a man on foot was no man at all. Like the Longhorn, the mustang has been virtually bred out of existence. But mustang horses will always symbolize western frontiers, long trails of Longhorn herds, seas of pristine grass, and men riding free in a free land.—J. Frank Dobie." (From Alexander Phimister Proctor, *Sculptor in Buckskin: An Autobiography* [Norman, University of Oklahoma Press, 1971])

worried about how fast the wild horse was disappearing and decided that they would see that at least a few were preserved. They began a search-and-save operation early in the 1920's.

The most important individual of the group, particularly because of his ability to attract publicity, was Bob Brislawn, of Oshoto, Crook County, Wyoming. He and his brother, Ferdinand, decided that they must save a few true Spanish mustangs. It was not an easy task, and it demanded a lot of travel and riding the back country of several Western states before they succeeded. They obtained most of the horses from Indian tribes, Crows, Cheyennes, Shoshones, and Utes. They received a real break when a few small Indian pony mares were brought into Crook County from the Crow Reservation in Montana by Charles and Dick Williams in 1925. A little later they picked up two more Spanish mustang mares in New Mexico and a small buckskin mare from across the border in Old Mexico. The remaining base stock they obtained from Monty Holbrook. Holbrook, a mustanger all his life, had captured a three-year-old stallion in 1927 in the Book Cliffs area of Emory County, Utah. He also had some pure mustang mares and, he had bred this stallion to his mares. The Brislawns obtained five grullo and one buckskin mare from Monty Holbrook, bred to the feral buckskin stallion that Holbrook had captured.

Two of the foundation stallions—foundation for the Brislawns and the Spanish Mustang Registry—came from the Holbrook purchase. One stallion was a grullo called Buckshot, who was registered number 1 in the registry, and Ute, an orange dun, who was registered number 2. The third horse registered was Yellow Fox. He was obtained by Bob Brislawn from the Cheyennes.

Bob Brislawn spent almost forty years trying to arouse interest in pure Spanish mustangs so that a breed organization could be founded to preserve them for posterity. Finally in 1957 his labors bore fruit, and the Spanish Mustang Registry, Inc., was established at Sundance, Crook County, Wyoming. Brislawn received valuable assistance from Lawrence P. Richards, of the University of Illinois, who was also interested in mustangs. The stated purpose of this group is to preserve the last remnants of the pure Spanish mustang by registering the finer and better-authenticated animals.

The authentication is a problem. Over the years Bob Brislawn collected and evolved a strict list of requirements about just what a pure Spanish mustang was—and was not. Each horse he obtained met those requirements. These characteristics were listed as the description of a true Spanish mustang, and only a few other breeders had such horses. The very scarcity of the horses caused problems in the association.

Recently another group devoted to the mustang was organized in Colorado. It was named the Spanish Barb Mustang Breeders Association. In a fashion similar to several other Western breed organizations the group registers horses in three divisions, or categories: permanent, temporary, and appendix—the idea being to allow all eligible mustangs to register and to prove themselves. If it is found out that they are not true mustangs, they will not be advanced to the permanent registry.

Actually, there is not much difference in the objectives of the two groups—only in the way they aim to achieve their goals. The Spanish Barb group are trying to breed typical breed characteristics and to eliminate atypical conformation. They feel that this is the only satisfactory way to proceed, since there are not enough true mustangs available to create a registry.

Pegg Cash, secretary of the Spanish Barb Mustang Breeders Association, said in a letter to me dated March 16, 1973:

> This in a nutshell is the reason the SBMBA came into being, to allow those animals that deserve registration through proof of authenticity to receive certificate, and to keep those animals that prove unworthy of the breed from continuing to produce non-characteristics of the breed.

This point of view is in direct conflict with the stated objective

Above: Buckshot, number 1 in the Spanish Mustang registry. He is Bob Brislawn's foundation stallion. Below: Coche Tres, a Spanish Barb Mustang mare, bred and owned by Susan Banner, Calhan, Colorado.

of the Spanish Mustang Registry, which states flatly that "no attempt whatsoever will be made to improve the registered mustang."

There is yet one more horse registry that might well be placed with that of the mustang. It is the American Indian Horse Registry, which is dedicated to the preservation and upgrading of the American Indian horse. Originally the home of the registry was in Phoenix, although at present the office is in Apache Junction, Arizona. In the Manual of Regulation (1968) under "Purpose of the Organization," the following is found:

> Though it has long been only a dream, it has been the life ambition of the originator of the idea, that the American Indian horse should hold not only a place in history, but should also have and earn his place in the present and the future.

Larry J. Klutt, the founder, in a letter to the present author dated April 20, 1973, said when you get down to basics, the horses in his registry are the grade horses in America, the American family horse.

◆§ *Glamour Boys, the Palomino*

When one first sees a golden horse with snow-white mane and tail, he generally exclaims, "That is the most beautiful animal I have ever seen." And truly these tawny *caballos* flecked here and there with spots of gold, accentuated by their ivory points and black hoofs and eyes, are one of the bonniest sights granted to man.

Much has been said and written about Palominos, and much that has been said and written will not stand investigation. A few examples will not be out of place. It has been said that Palominos have been a breed since the earliest times. Desirable they have always been, and are—a breed they probably never have been. Even if one had a breed of horses called Palominos, that would not stop the same color from appearing on other horses who would also be called Palominos. Tales have been told which claim that Queen

Orren Mixer's famous painting of a Palomino stallion.

Isabella of Spain sent a stallion and five mares from the *Remuda Real* to Mexico—Her most Catholic Majesty died in 1502; Mexico was discovered in 1519. It has been said that Cortés brought over Palominos with him for his conquest of Mexico, even that his own *castaño zaíno* which died at San Juan de Ulúa, or his *Morzillo* who became a god, was a Palomino. None were Palominos. Some persons have claimed that Palominos were never sold, but they were, as often as their owners wished to sell. Their name has been traced to bottles of wine, pigeon dung, soiled chemises, and California soldiers. All these and many other tales and superstitions have been related; however, they should not be held against the "Golden Ones.

Today the Palomino is probably the most popular color that graces the body of a horse. Like bays and browns and sorrels, it has many variations. Palominos run from the color of a well-seasoned and polished Osage orange bow to the light yellow color found in a "half-and-half" cream carton. Their manes and tails run from a silver color through ivory to snow white. Some horses are dappled; some are not. It is common practice to call any yellow horse which does not have a black mane and tail a Palomino, and in the long run, common usage seems to carry weight. Men have called them "Claybanks"; Mexicans sometimes call them *Cobre, Andaluz, Bayo Claros,* or according to a few people, even *Palomino.* Dick Halliday, who is credited with starting the first registry and was the original secretary of the Palomino Horse Association, in the 1938 *Palomino Parade* defines the ideal coloring of a Palomino: "In full coat, the body is of burnished gold . . . [the] head should carry a full length blaze. White stockings, halfway between knee and hoof, on the front legs, and on the hind legs, white stockings again, halfway to the hocks. And these hoofs they spring from should be . . . black." Halliday's markings are those generally accepted as ideal, and most Palomino breeders are trying with varying degrees of success to produce them on their horses.

Booger Bear, a champion Palomino stock-horse stallion, owned by Jack Bridges, Glen Rose, Texas.

When one begins talking about the ideal conformation of a Palomino, the real argument starts. The Palomino Horse Breeders of America, the most successful Palomino horse association, is attacking the problem by admitting that the Palomino is today, and probably always has been, only a color and by trying to make that color consistent in several different types which the breeders are using. It is questionable whether the result can be called a breed, since there is no common breed characteristic except color; but it is the only practical step to be taken under the circumstances.

Despite a common impression, it is not possible to buy Palominos and, simply because they are registered, expect them to beget Palominos. Trying to establish a breed, on a color seemingly a heterozygous characteristic, is a difficult proposition, as more than one fresh Palomino enthusiast has discovered. Practically any color will appear, with the chances somewhat in favor of albino. Perhaps the greatest success has thus far been obtained by crossing sorrel and Palomino. This prevents the appearance of white skin, eyes, and hoofs, the next step to albinoism. On the Pacific coast the original effort to establish a breed was done to a large extent by the infusion of Arabian blood. Dick Halliday said that this is the only sure method of getting Palominos. To a certain extent, this means is acceptable, but if used very much, it creates a horse whose conformation is not the type which other sections of the country want in their favorite mounts. Texas breeders have used more sorrel Quarter mares in their breeding program. One investigation made of the various breeders in the West showed that in many respects this was the most successful, as the resulting horse was one which most Western riders find suitable to their needs, regardless of the color which resulted.

There are three names which stand out whenever one starts talking about the history of Palomino breeding in the United States: Coke Roberds, who owned that incomparable Palomino Quarter Horse, Old Fred; W. B. Mitchell and his famous Palomino

Sappho; and Dwight Murphy, owner of Del Rey. Roberds' home range was Hayden, Colorado; the Mitchell ranch was at Marfa, Texas; and Murphy owned Rancho San Fernando Rey in California.

Coke Roberds had always been a horseman, as his father was before him. Seldom is it the privilege of one man to own two truly great horses, but such was the privilege of Roberds. As he himself once said, "A truly great horse is a rare accident, and the laws governing accidents have not yet been revealed to man." Coke Roberds was in his lifetime the owner of Peter McCue and Old Fred, names which will never grow old while there are horsemen in the Southwest.

Old Fred was the greatest Palomino that ever lived in Colorado and, some think, probably in the whole United States. His blood is in practically every leading Palomino band in the country.

The story of how Roberds happened to get Old Fred is interesting. One day, soon after the turn of the century, when Coke was on his way to Steamboat Springs, he had to draw up to let a freighter plod by. His eyes rested on a wheel horse of the outfit, and he straightened. Giving the reins to his wife, he said, "I must have that horse."

Going over to the freighter, he asked, "What do you want for that dun horse?"

"Don't want to sell him."

"What do you want him for?"

"To work."

"I'll give you a better work horse." Time and the bargaining went on, but Coke did not budge until he had bought Old Fred from out of the traces.

Old Fred was not a small horse. He stood sixteen hands and weighed 1,440 pounds. However, he was a Palomino, and except for his size, sported ideal Quarter Horse conformation. He was just a large edition. Although he did not have a racing record, all of his colts were runners. Some horsemen hold that Sheik was the best descendant of Old Fred. Sheik was out of an Old Fred filly by Peter McCue. He couldn't miss with that heritage. Sheik was owned by the Matador Land and Cattle Company and was one of the great Palomino studs of the Southwest. Squaw was the fastest Old Fred filly. Although run countless times, she was seldom beaten. Another great descendant of Old Fred was Plaudit. His dam goes back to Old Fred, while his sire was King Plaudit by Plaudit, who won the Kentucky Derby. One of his greatest sons was that fast and beautiful Palomino, Question Mark.

A characteristic which really impressed one on seeing Old Fred, even more than his beautiful color, was his tremendous musculature. Coke says when speaking of his forelegs that they "forked like an oak tree." He had such good withers that one could throw a saddle on Old Fred and ride him without a cinch. Absolute assurance of his pedigree unfortunately cannot be ascertained. However, it appears that he was by Black Ball by Rondo and out of a mare by John Crowder.

An incident similar to the one told about Roberds started W. B. Mitchell of Marfa, Texas, breeding Palominos some fifty years ago. One day while he was riding, alongside a fence he saw a little yellow colt frisking around his mammy. After watching him for a while, Mitchell decided that he had to have him. He found the owner of the colt, bought him, and named him Sappho. Sappho's dam had a strong infusion of Arabian blood, and his sire was half Quarter Horse and half Thoroughbred.

Sappho grew up into a stallion, gentle in disposition and unusually intelligent. His colts possessed these same qualities to a remarkable degree; they could be trained to do anything. Many of Sappho's sons and grandsons went to California. At one time Mitchell sold a carload of mares to go to Santa Barbara. Sappho's progeny stood in almost all of the Southwestern states, not to mention Idaho, Wyoming, Indiana, and Kentucky.

Dwight Murphy of Santa Barbara, California, was probably the first man of combined ability and means seriously to enter the Palomino game in California. He obtained a yearling stud at Ventura which he called Del Rey, who was sired by Swedish King, sired by El Rey de Santa Anita, the old racer owned by Lucky Baldwin. He had difficulty in getting a group of Palomino mares until he heard that Mitchell, of Marfa, Texas, was raising Palominos. El Rey de los Reyes was the best son of Del Rey. He was considered by many the most beautiful Palomino alive. Hijo del Rey and Conejo were two other beautiful Murphy studs.

It must be said that some of the earliest Palomino breeders did use Thoroughbred blood. Mitchell used a dark-red government stud, Lantados, on Palomino mares and got satisfactory results. This horse was a son of Pillory, the chestnut stallion which won the Preakness. Roberds had some Thoroughbred blood, although for the most part he used short-horse blood. Dwight Murphy's foundation blood traced to the Thoroughbred, and he also used two Arab stallions.

Although the move to organize Palomino breeders started in the 1930's, it was not until 1937 that the first studbook appeared. It was called *The Palomino Parade*. It was compiled and edited by Dick Halliday, and a new one appeared yearly until 1945, with about three thousand horses being registered. After 1945 the Palomino Horse Association, at least as operated by Dick Halliday, ceased to function.

The Palomino Horse Breeders Association gradually took over leadership of the Palomino registry. It was chartered in the state of Texas in 1946. The purpose of the organization, in their own words was "to provide for the registration, purity of blood, and improvement in the breeding of Palomino horses, and keep, maintain, and publish, in suitable form the history, record, and pedigrees thereof."

The early ramrod of the Texas organization was Dr. Arthur

Zappe, of Mineral Wells, assisted by such ranchers as Roy C. Davis, W. B. Mitchell, and Howard B. Cox. They also established a monthly magazine called the *Palomino Parade*, which has appeared regularly for over thirty years. Throughout the years the association has kept close tabs on the horse activities by approving shows, drawing up rules, approving judges, and certifying various state organizations affiliated with the P.H.B.A. There are over thirty such affiliated associations today. When Dr. Zappe retired, his place was ably taken by a new executive secretary, Melba Lee Spivey.

The P.H.B.A. requires proof of breeding for registration. The sire or the dam of the horse to be registered must be registered in the P.H.B.A. or one of the following studbooks: *The Quarter Horse*, *The Thoroughbred*, *The American Saddle Horse*, *The Arabian Horse*, *The Tennessee Walking Horse*, *The American Remount*, or *the United States Trotting Horse.* Originally they also accepted horses registered in the Palomino Horse Association. The various shows and performance classes are divided into type-classes: Quarter Horse, Saddle-bred, jumpers, cutting horses, Western Pleasure, and several others. Mares or stallions with albino or piebald sires or dams cannot be registered.

An interesting survey showed that of the top show horses 70 per cent had Palomino sires and 18 per cent had sorrel or chestnut sires. Of the breeds outside the P.H.B.A. about 90 per cent of the sires were Quarter Horses, and 10 per cent were American Saddle Horses. Of the dams, 44 per cent were Palomino, and 30 per cent were sorrel. As a whole about 27 per cent were the result of Palomino-to-Palomino mating. The association feels that this indicates some progress over the years in achieving a measure of fixation of color from the top horses.

The Palomino caught the public fancy and became one of the most popular of all show horses. Although Palominos are not as yet a definite breed, insofar as having a particular utility or conformation, and even though they cannot be bred true to color with

A Palomino stallion. (Courtesy John A. Stryker)

any real consistency, the interest that has been aroused, the numbers of people trying to raise them, and the associations that have sprung up in the various states will tend to standardize this horse.

❧ *Peculiar Spotted Ponies, Appaloosas*

"These peculiar spotted ponies come from either the Umatilla camp- or Nez Percé stock—they ain't bred by no Indians east of the Rockies." Charley Russell spoke these words. He was fascinated by multicolored horses and always noted when an Indian was riding an unusually colorful horse. Russell's outlook is still shared by Western men. Anyone who has seen the infinite colorations that appear on the wild mustang cannot but notice and admire the unusual color formation. Today there are several strains of color which are being selectively bred. They include the Palominos, Appaloosas, Pintos, leopard-spotted Colorado Rangers, Albinos, duns, and grullos. Perhaps the most striking coat to appear on a horse is that of the Appaloosa, the beautiful horse which generally has a spotted rump. Horses of this sort are also known by the color names polka dot, raindrop, leopard spot, dollar spot, harlequin, spotted rump, speckled tails, and several other variations found in different localities.

An Appaloosa is easy to identify. Each one has characteristics that set him apart from other breeds. The skin is mottled with an irregular spotting of black and white. It is especially noticeable about the nostrils. The eyes are encircled with white like a human's, and the hoofs are striped vertically with black and white. Naturally the coat patterns vary radically. While the white on the loins is not universal, it is certainly preferred, especially when it is accompanied by spots, round or egg shaped and varying in size from specks to four inches in diameter. One thing for sure—no two horses are alike.

Like the old coyote duns, Appaloosas are the subjects of many

An Appaloosa. (Courtesy *The Cattleman*)

stories connected with their speed and endurance, and many famous persons have ridden them by preference. Bill Williams, an old mountain man whose name is written large in the history of the North and the West, rode an Appaloosa all over the country and claimed that his horse was tougher than a Missouri mule. If the horse was, he certainly made a most fitting companion for his master. Bill rode this cropped-eared spotted horse until the day he was found under a tree with the feathers of an arrow protruding from his chest. There is another well-remembered spotted horse who has almost as many legends connected with him as the famous "White Steed of the Prairies." He was called the "Fan-tailed Appaloosa" and ranged the Crow country in the eighties, always moving, never caught.

There are various explanations of the name "Appaloosa," although the correct derivation is from "Palouse." The horse seems to have been found in the Palouse Indian country in rather large numbers, particularly among the Nez Percé Indians. At first it was called "a Palouse" horse, which was soon contracted to "Apalouse," which is now spelled "Appaloosa." There is a Spanish noun "pelusa," from which several writers have hastily decided the word came; still others contend the spelling is "Appaluchi" and connect the name with the Appalachian Mountains. About the only source which has been overlooked is the statement by J. J. Hooper's incomparable character, Simon Slugg, who said, "Right around thar war where I ketcht the monstrousest, mos audaciousest appaloosas cat [fish] the week before that ever comed outen the Tappapoosy."

Oregon has for many years been one of the most prominent states breeding Appaloosas. Old Painter, a famous snow-white steed with black spots on his hips, was perhaps more responsible for Oregon's dominant position than any other single horse. Old Painter's most famous son was called Young Painter and was out of a black mare from the Prairie City Rodeo bucking string. Lee and Christy Le Roy used two of his colts in their trick riding and roping, while many other colts found their way into cattle outfits, riding stables, and circuses.

The Painter strain of horses was developed by Claude I. Thompson of Moro, Oregon, who became president of the Appaloosa Horse Club, Inc. Thompson's basic stock was Painter blood crossed with pintos, leopard-spotted mares, and a few Palominos. With the idea of improving his horses, he soon bought a registered Arabian stallion, a sorrel with a light mane and tail. With this stallion he produced some unusual horses, particularly Painter III, a leopard-spotted colt; Arabian Girl, a pinto-spotted filly; and Cremolyn, a registered Palomino. Crossing his Arabian stallion with its sorrel coat on his multicolored mares, he secured some startling results.

Mr. Thompson and Dr. Haines soon got together several other interested persons and the Appaloosa Horse Club, Inc., was formed. Mr. Thompson was elected president; Dr. Francis Haines was made historian; Ernest A. Kuck was elected vice-president; and Faye Thompson was made secretary-treasurer. The purpose of the organization as stated in the by-laws was to improve and standardize the strain of leopard-spotted horses known as Appaloosas, to create a suitable foundation stock, and to collect records and historical data. In the beginning, to be eligible for registration, a horse had to have a white rump starred with round or oblong spots and also flesh-colored spots around the eyes, nose, and genital organs. A stallion had to be dominantly leopard-spotted with a white rump covered with dark spots. Any base color, such as grey, bay, chestnut, cream, white, roan, and so on, was acceptable. Mares that threw spotted colts were recorded by the group but not registered.

The distinctive traits listed by the organization as desirable in the Appaloosa horse are as follows: Any base color is permissible as long as leopard spots are evident. Although most breeders prefer a blanket on the hips, not all Appaloosas have them. It is suggested that the height should not be less than fourteen hands, two inches, or over sixteen hands. The weight should fall between 900 and 1,100 pounds. The head should be broad with a full forehead, the eyes black or hazel with white around the edges, and the feet parti-colored. The other desirable qualities listed are those found in any good saddle animal.

The purpose of the Appaloosa Horse Club has been to preserve, improve, and standardize Appaloosas, the leopard-spotted horse that is termed an Appaloosa in the Northwest. The association has registered horses and collected historical data about the breed. As the new breed grew, it became necessary to have a permanent executive secretary, and George B. Hatley was selected. He proved to be a popular choice and an efficient secretary. Under his direction the association has made remarkable growth. Advertisements are carried in appropriate journals, eleven studbooks have appeared between 1950 and 1972, and an excellent book, *The Appaloosa Horse*, by Dr. Francis Haines, Robert L. Peckinpah, and George B. Hatley, was published in 1950. In 1963, Dr. Haines brought out the best source yet published on the history and background of the breed: *Appaloosa: The Spotted Horse in Art and History.*

Registration has grown steadily, with almost 200,000 registered since 1938. The "Appaloosa Horse News" has also grown from a one-page mimeographed sheet to the present slick magazine going to thousands of interested horsemen. Appaloosa horses are also being run, in the Quarter Horse fashion, in several states. Since 1970 all Appaloosa foals must have their sires and dams registered or identified by the Appaloosa Horse Club. The Appaloosa horse is a going concern.

Peculiarly enough, there is another group of individuals who are raising horses, most of which many stockmen would call Appaloosa, or leopard-spotted, horses, but they are registered as Colorado Rangers, the group preferring to call them leopard-spotted horses. The Wine Glass Stables at Boulder, Colorado, owned by

K. K. Parsons, was the home of the Colorado Rangers in the 1940's. Parsons claimed that Colorado Rangers were American-bred Morocco Barbs derived from foundation stock imported from Andalusia, Spain, by a former governor of Colorado. The body of a Colorado Ranger according to his description, is pink-skinned and covered with a milk-white coat of silky hair marked like the coat of the desert leopard, with a myriad number of small jet-black dots varying in size from those of a quarter to those of a silver dollar. The leopard markings are embossed on the white coat and can be felt by touch of hand. Twenty-one different contrasting color combinations are recorded in the first three volumes of their studbook. While the registry association makes no requirement concerning a definite color pattern, emphasis has constantly been placed on odd, flashy coloration.

The name Colorado Ranger implies that the horses are allowed to range the year around. Parsons allowed his stallions to run with the mares on open ranges and claimed nearly a 100 per cent colt crop. Mature Rangers, produced on grass and a minimum of grain, weigh from 1,000 to sometimes well over 1,300 pounds, and run in stature from fifteen hands, three inches, to sixteen hands, two inches.

The following account of the formation of the Colorado Ranger breed is taken from a letter from Parsons to me dated June 7, 1939:

> The history of the breed traces back to the two stallions presented to General Grant by the Sultan of Turkey [one was Linden Tree and the other Leopard, but curiously enough Leopard was the Arabian and Linden the Barb]. Mares sired by these stallions were bred to a pure Barb stallion imported into Milam, Tennessee, from Morocco. The offspring resulting from this cross were then crossed with a Barb Leopard stallion imported into Colorado by Governor Shoup, from Andalusia, Spain. The famous Hambletonian Aegondal 42357 was used on many of the best Leopard mares of the breed a decade ago. Being a chestnut, his

cross-bred offspring were born with startling leopard coats. Till very recent years all members of the breed were retained within the State. Officers and directors of the Colorado Horse Association, Inc., felt that until the breed had become thirty years old, all breeding stock should be kept available.

The story of the Colorado Rangers is given a little different emphasis here and there by John E. Morris, a member of the Board of Directors of the Colorado Ranger Horse Association in an article in May, 1972, issue of the *Western Horseman*. According to Morris, in 1934 a Colorado rancher named Mike Ruby showed two of his horses at the Denver National Western Livestock Show. They were both leopard-spotted and drew a lot of attention. As a result in 1937 the name Colorado Ranger was selected to signify the Colorado horses bred on the open range. An organization was formed, and Mike Ruby became its first president.

The normal time necessary to build up a breed had already been expended by Ruby, who had wisely kept the pedigrees of all his horses, and he had also instituted a grading system for his Rangers. The *Western Horseman* article states: "There was no search for foundation stallions and mares; they were already in existence and had the characteristics and fixity of a breed."

Good using horses were greatly in demand in Colorado at the turn of the century, and the attempt to fill this demand resulted in the first Colorado Rangers. Several Colorado ranchers got together and selected A. C. Whipple, of Kit Carson County, to go to General L. W. Colby's ranch in Nebraska to obtain some of his horses. Colby's horses traced to a couple of Turkish horses that Ulysses S. Grant had obtained in 1878 from Sultan Abdul Hamid II. One was called Leopard and was supposedly an Arabian. The other, Linden Tree, was a Barb. Whipple was successful and returned to Colorado.

The horses he obtained included a band of mares and one stallion. The stallion, named Tony, was a double grandson of Leopard. Breeding these mares and Tony resulted in horses whose coats were kaleidoscopic. Then W. R. Thompson, of Kirk, obtained a Barb stallion named Spotte, who had come from Algeria. Thompson lived not far from Whipple, and Spotte was bred to Whipple mares. Again odd, barbaric color patterns occurred. Then Mike Ruby bought a stallion named Max from Oliver Shoupe, governor of Colorado, and crossed him on some mares he obtained from Whipple. The stallion, Max, had been sired by the Waldron Leopard and was out of an Arabian mare. It was the offspring from Max and Whipple mares that Ruby took to Denver in 1934.

Ruby died about the time of World War II, and the organization lay dormant. Membership had been limited to fifty. In 1967 a new president was selected, and unlimited memberships became available. Interest grew, and soon there were three hundred members, and registration reached four figures.

There is no color requirement for eligibility at the present time, although to be registered the horse must trace directly to one of the two foundation stallions, Patches, registered number 1, or Max, registered number 2. Patches was a Whipple stallion sired by Tony. Outcrosses are accepted, if they are registered Quarter Horses, Thoroughbreds, Saddle Horses, Tennessee Walking Horses, Appaloosa Horses, Morgan Horses, or Standardbred Horses. The association hopes to close the books to all outcrosses in 1978, which will mark the centennial of the importation of the foundation sires, Leopard and Linden Tree.

A rather consistent program of line breeding and the naturally dominant character of spotted horses to throw spotted colts have made the Colorado Rangers good breeders of their kind. Their high withers, short back, and snappy action, combined with their spectacular coloration, all go to making a horse which a great many people find desirable.

◆§ *Painted Ponies, Pintos*

There is no equine as familiar in popular scenes of the West as the two-toned paint pony. To many people this horse represents the romance of Indians and rodeos, cowboys, and mustangs. No Western show would be complete without a colorful Pinto horse to carry the flag.

Pinto horses have been common throughout the known history of the world. They are seen on the art works of the early Chinese. They have been represented in the tombs of Thebes. They are pictured hitched to the chariots of ancient Egypt. Jacob in the Bible expressed a preference for Pintos; they are common in pictures of medieval Europe; and two of Cortés' original importations were multicolored horses.

Pintos have not always been universally popular with North American horsemen. Perhaps one of the main reasons is that for many years few breed organizations allowed spotted horses to be registered. To say that a Pinto horse, because of his color, is worthless, is like saying that a two-toned automobile is worthless. It is what lies beneath the color that makes the horse, as it does the car. It may be true that in the past little attention was paid to the breeding of Pintos and that many were therefore inferior, but that would not be the fault of the color. There are other colors, for example bays and sorrels, on equally worthless horses.

Aime Tschiffely rode a paint horse from Buenos Aires, Argentina, to Washington, D.C. That would hardly indicate lack of endurance or worthlessness. On the contrary, his was one of the most astounding rides ever completed and is a marvelous recommendation for the *Criollo* horse of South America.

Some believe that the Pinto is the result of the indiscriminate breeding that took place between horses in the wild mustang bands. If spotted horses were the result of inbreeding, then the color of the Thoroughbred and other modern breeds would be kaleidoscopic. Unless at least one horse with Pinto blood was included among the originators of the wild herds, there would have been no Pintos.

In 1941 a Californian, George Glendenning, organized the Pinto Horse Society, whose objective was to allow the breeders and admirers of the spotted horses to work together in gaining recognition for the merits of the horse and to improve the Pinto horse in the United States.

The nine principle objectives were:

1. To study and perpetuate the spotted horses and ponies as a part of our Western heritage.
2. To keep an accurate record of horses registered with the Society.
3. To publish a yearbook devoted to the Pinto horse.
4. To educate the public to the value of Pintos and disperse information and publicity.
5. To ascertain the scientific breeding laws for the production of better Pintos.
6. To encourage the inclusion of suitable classifications for Pintos in horse shows.
7. To establish rules for judging of Pintos in competitive events and to standardize them throughout the country.
8. To assist and encourage breeders.
9. To compile a free booklet listing registered breeders.

The original Pinto organization was active for a few years but by 1947, owing to World War II as much as anything, had slipped into inactivity. Another factor causing difficulty was the common trouble of all breeds founded on color. Different breeders wanted to use their Pintos for different purposes. Often these activities had little in common, and the desirable conformation was dissimilar. When any attempt was made to promote one type over any of the others, dissension occurred among the breeders.

In December, 1954, Kay Heikens published an article in the *Western Horseman* magazine, "The Plight of the Pinto." A lot of

Mister J Bar, a sorrel *overo* Pinto. He is a national champion roping horse, registered with the American Paint Horse Association, owned by Junior Robertson, Waurika, Oklahoma. (Photograph by Margie Spence)

interest was generated by this article, especially among the better Pinto breeders. The outgrowth was a new organization, the Pinto Horse Association of America. It was incorporated in New Jersey by a small but dedicated group of serious Pinto breeders. The new registry provided for a tentative and a permanent studbook, as well as for a group of solid-colored breeding stock. The organization also maintains a pony division for registering Pintos under fourteen hands.

There is still another registry for Pintos, the American Paint Horse Association. It was started by thirty Paint Horse breeders from Texas and Oklahoma, who first met in Gainesville, Texas, in 1962. Their aim was to register, promote, and keep records of stock-horse, or Quarter-type, Pintos. They planned to follow the Quarter Horse Association as a guide to their activities wherever appropriate. Apparently it was a good direction to follow, for they have been remarkably successful. The first year they registered 250 horses. Within ten years they had over 20,000 registered Paint Horses, located in all fifty states and Canada. In 1973 I received an informative letter from the Paint Horse Association's able executive secretary, Sam Ed Spence, which is worth quoting. The letter was dated March 5:

> Concerning our Stud Books, the Number 1 Volume 1 came out in 1965 with horses listed through #2,599. Stud Book #2 was published in 1968 and included horses from #2,600 through #6,000. The #3 Volume was published in 1970 and included horses from #6,001 through #12,000. The #4 Stud Book is in the mill now and will carry through #18,000.
>
> The Paint Horse Journal began publication in November, 1966. It is a bi-monthly publication and for the past three years we have also published a Newsletter between issues of the Journal.
>
> The Association is now publishing its first racing chart book which will include the official charts of all Paint Horse races from 1966 through 1972. We began with the 1966 races since this was the first year that we

had an official performance department and started keeping records on races and shows.

I am enclosing a copy of our 1973 convention report which will give you a progress report on the Association since its beginning in 1962. Am also sending along a copy of our Paint Horse brochure which will give you a brief run-down on the history of the Association.

Mr. Denhardt, I surely appreciate the opportunity to beat the Paint Horse drum. Like all the other light horse breeds with which you have been so familiar over the years, we boast of our horses versatility, disposition and tractability. But as you well know, such traits depend upon the individual horses, regardless of breed.

I do think, however, that our association and its various activities has something to offer the individual owner or breeder that might not possibly be available in other registries. Comparatively speaking, we are still a small registry and we make a special effort to be as informal, personal and accommodating as we can to our membership and still uphold the ideals of a breed organization. Our basic philosophy here is that people should enjoy their horses and enjoy the association with their fellow horsemen. Our common bond here, of course, is this spotted horse which because of his individuality and color pattern affords his owner a certain distinction.

The American Paint Horse Association has an official Rule Book . . . tailored after the American Quarter Horse Association. One primary difference is that we do not have Honor Roll horses, except in the Youth Division, but have instead a National Show each year whereby all horses are eligible to compete on a three go-round basis in order to select National Champions in each sex, age division and performance event. Of course, we also have Supreme Champions, APHA Champions, Register of Merit awards and Superior Horse Awards very much the same as does the AQHA.

Both Pinto associations distinguish their horses in the South American manner, describing the horses as *tobianos* or *overos*. The tobiano is what the average Westerner calls a "pinto," since the horse is not a roan, and the spots are clear cut and normally large. Taking white as the basic color, in the tobiano it originates on the

Above: Sun Cloud, a *tobiano* Pinto stallion, owned by Fleet Medlin, Hayward, California. Below: Nellie Pinto, an *overo*, owned by D. C. Jackling, Woodside, California.

back and rump and works down in large, smooth blocks. The overo is often a roan horse, and, with white again the basic color, it is found to extend upward from the belly with the darker color appearing in many small ragged patches. The overo generally has a bald or mottled face, while the tobiano normally has face markings similar to that of the average horse. The overo may have a mane and tail of two colors, but they are generally mixed. If the tobiano has two colors in his mane or tail, they are always distinct and separate, and the color line is clearly broken. The tobiano is more common in North America and the overo more common in South America. In fact, the tobiano, according to Roberto C. Dowdall, is of late origin in the *Criollo* of South America.

Albino, Dun, and Grullo

The Albino, a pure-white horse, has always attracted attention wherever it is seen. White horses have ever been popular. Gods, kings, generals, and statesmen are commonly depicted riding pure-white stallions. Today a white horse is considered a favorite in the parade, the circus, the show ring, and as a "high-schooled" trick horse. He has also found a place in the rodeo.

Twenty-five miles north of Stuart, Nebraska, is the old ranch home of Cal Thompson, who was responsible for the formation of the American Albino Horse Club. Thompson began breeding white horses in 1919. His interest in their origin and improvement did much to popularize the horse. The foundation sire of the Thompson string was called Old King, foaled in Ohio in 1906. He stood fifteen hands, two inches, and weighed 1,200 pounds. His skin was spotless pink, his hair milky white, with long silky mane and tail, and his eyes were dark brown. Old King came to Stuart, Nebraska, in 1918 and was used for ten years, dying of swamp fever in 1928.

Measurements for the perfect Albino horse as set by the Albino

Snow Chief, a registered Albino stallion.

Club are: height—fifteen hands, two inches; weight—1,150 pounds; color—snow white; skin—pink; hoofs—white; nineteen and one-half inches around muzzle at top of mouth; seventeen inches from top of mouth to base of ear; nine inches wide between the eyes; thirty-three-inch throatlatch; thirty-five inches from poll to center of wither; twenty-six-inch back (from center of withers to hip joint); eighty-two-inch heart girth; thirty-two inches from girth to ground; and eighty-inch flank girth.

In a letter to me dated March 19, 1973, Ruth Thompson White, who has been the secretary of the Albino Horse Association since its inception in 1937, said that she has had great satisfaction in her work, for she knows that the records are being kept accurately.

The greatest change that has occurred in the last thirty years is the separation of the registry into two sections: the American Whites and the American Cremes. Until September, 1949, only pure-white horses could be registered. However, in 1942, Dr. W. S. Castle, a college professor, wrote an article that changed the thinking of the Albino breeders. They began to register horses that shaded from ivory through cream, as well as pure whites. Light or blue eyes were also acceptable. Dr. Castle pointed out in his article that type A Albino horses, when crossed on a sorrel, would produce nearly 100 per cent Palominos; type B Albinos would produce 50 per cent palominos and 50 per cent buckskins; type C would produce mostly buckskins; and type D Albinos, the dominant whites, would produce 50 per cent white and 50 per cent cremes. So every attempt was made to classify all Albinos by their letter type.

From 1939 to 1949 a national show was held each year at the Thompsons' White Horse Ranch at Naper, Nebraska. The years 1949 to 1963 were a period of little activity for the breed, and the last White Horse roundup was held at the White Horse Ranch on June 14, 1963. Cal Thompson died not long afterward.

Later that year a reorganization meeting was held, and all new officers (except for Ruth Thompson White) were elected. The

Thompson's National Velvet (left) and Wyoming Lady, registered American White Horses bred by Ruth Thompson White and owned by Charles C. and Bertha R. Ludwin, West Alexander, Pennsylvania.

American Albino Horse Club was moved to Oregon and reincorporated under the name American Albino Association, Inc. The greatest change was the one that allowed many more horses to be registered. A new division was set up, called the National Recording Club. In it types A, B, and C would be registered. For the ultimate horse, type D, the dominant white, a tentative and then a permanent registry was established. In November, 1970, the board of directors voted to place the cremes in a section of their own in the registry and to call them the American Creme Horse. The final change was made in 1970, when all dominant white horses were registered under the name American White Horse. The word albino had to be eliminated, because a true albino is an animal which lacks pigmentation and consequently has oversensitive skin and eyes.

The American White Horse (the old Albino) may be of any acceptable type, and the registry follows the pattern of other color registries, listing the main breeds accepted, such as Quarter Horse and Thoroughbred.

There are other Western types which have separate classes in stock-horse shows. These are duns and grullos (or grullas), the former being the most common. The dun horse, often called coyote dun or zebra dun, depending on his coloration, is generally regarded as a tough, all-around cow horse. Most cattlemen will say that they never saw a poor dun. If all duns are as good as their reputation, then certainly there is no horse alive with the endurance, cow sense, strength, and stamina of this animal. J. Frank Dobie said, "As the ranchers of the Southwest know, no horse hardier or better adapted to range work than the Coyote Dun has ever been ridden by a *vaquero*." Dobie's sentiments are shared by ranchers all over the West. Duns range in color from a dirty yellow or lemon through tan and dark buff to a sulphur or golden bronze. They normally have black points. Often they have a black stripe down their backs and lateral black stripes rather high on their

A dun (left) and a strawberry roan. (Courtesy *The Cattleman*)

legs and often over the withers. There also are red duns—sorrels with a red line down their backs.

The grullo is mouse-colored. The main part of the body is almost slate-colored, while the points are always black. Zebra markings on the legs and the stripe down the back are the rule. Grullos vary from a salt-and-pepper color to almost a blue or mauve. They too have the reputation of being extremely tough and hardy. Incidentally, their name, from the Spanish, means "crane."

Today there are two organizations dedicated to the dun and the buckskin horse, the American Buckskin Registry Association and the International Buckskin Horse Association, Inc. The latter organization, under the leadership of the present executive secretary, Richard Kurzeja, is the most active. It is industriously promoting the buckskin, dun, and grullo horse.

The first registry was organized several decades ago in San Francisco. It failed to gain wide support, although it did much to popularize and promote study of dun horses. It became inactive in the 1960's. Then another registry appeared, this time in northern California, under the name American Buckskin Registry Association, with headquarters in Anderson. After a few years there was a schism, and still another dun registry was organized, the International Buckskin Registry Association. For a few years both organizations, working almost side by side, had problems. Meanwhile, in the Middle West there were a group of active buckskin horse breeders. Not being satisfied with either organization in California, they organized another club, the International Buckskin Horse Association. As its secretary wrote in a letter to me dated March 29, 1973: "The I.B.H.A. was formed and incorporated and money was invested to establish a proper organization for registration of horses with the dun factor." This new organization absorbed the older International Buckskin Registry Association, and Indiana became the home of the new group.

Almost immediately strong support arose for the new registry.

Today there are affiliated clubs in seven states and registered horses in forty-nine states. It might be of interest to know that their number-1 stallion in the registry is a registered dun Quarter Horse named Leo Reno, foaled in 1964. He has an enviable show and performance record, having competed with horses of both breeds.

In one of the official publications of the I.B.H.A. is the following description of the breed colors:

BUCKSKIN: The Buckskin has a body coat of a predominant shade of yellow, ranging from gold to nearly brown. Points (mane, tail, legs, etc.) are black or dark brown. On the true Buckskin the dorsal stripe, shoulder stripe and barring on the legs is always present. However, the dorsal stripe is *not necessary* for registration of the Buckskin.

DUN: The Dun differs from the "Buckskin" only in the respect that points may be of a lighter shade, even white. Usually the legs are only slightly darker than the body color. Duns with light manes and tails (with dorsal stripe) are eligible. The exact classification will be made in the I.B.H.A. office as to whether your horse is a Dun or a Buckskin, through the use of clear, colored photos.

GRULLA: The Grulla is the most rare, and the strongest phase of the "Dun Factor." Smokey blue or mouse colored, with black points. Do not confuse this color with a roan or grey horse. The Grulla (pronounced "Grew-yah") has no white hair mixed in with darker hair, as is seen in the roan or grey. The name Grulla comes from the Spanish, meaning "Blue Crane." Grulla hair is a solid mousey blue or slate color.

RED DUN: The Red Dun is just that—red. Body coat may vary from a yellow to a nearly flesh-color. Points are dark red. Dorsal stripe must be present.

Note: The Grulla, Red Dun and some shades of Dun must have the dorsal stripe to be eligible for registration. Dorsal stripe is not a requirement for the Buckskin.

The close distinctions among the various dun horses are necessary for registration. Western ranchers use a little looser terminology when identifying a buckskin, dun, or grullo. They have long been a favorite cow horse on the Western range.

13. In South America

◆ *Costeña, Chola, and Morochuca of Peru*

The region that belonged to the Viceroyalty of Peru was composed of many different kinds of terrain, pasture, and climate. As a result of this diversity the horses of Peru became acclimated to several varieties of feed and climate, and as time passed, underwent certain modifications. There arose three rather distinct types in the country, each developed by somewhat different use in very different environments. They are known today as the "*Costeña*," the "*Chola*," and the "*Morochuca*." The *Costeña* is the more stylish saddle horse and the *Chola* is the best horse for the ranches. The *Morochuca* is the small horse found in the high altitudes of the Andes.

The *Costeña* is naturally the horse in which most of the Peruvians take the greatest pride. A good deal of importance is attached to his style and to his smoothness of gait. The best gait of the *Costeña* is difficult to describe, but it is similar to an extended pace and has the animation of a rack. He can maintain this gait for long distances and at various speeds.

The height of the *Costeña* varies from thirteen hands, three inches, to fourteen hands, three inches. In the better examples, the head is carried rather high, the profile is straight, although in the old days many were somewhat Roman nosed. The forehead is flat and broad, and the ears are medium or small, flexible, and well separated. The eyes are bright, expressive, large, and correctly spaced. The mouth is dry and small, the muzzle delicate and clean.

The jaws are well placed. The neck is not extremely heavy but arched and muscular, especially at the base, and very flexible. The mane is full, fine, and abundant. The withers are not extremely high. The shoulders are sufficiently sloping, long and full, with the chest large and voluminous. The forearm and the cannon are clean and short. The knees are not too large. The pasterns are short and clean cut, with a slope of fifty to fifty-five degrees. The hoofs are strong, small, and well proportioned.

The back is broad, short, and flat, although many of the older mares become slightly sway-backed. The loin is short, broad, and well joined, and the ribs are well rounded. The *Costeña* generally measures from fifty-eight to sixty-one inches around the chest. The rear quarters are slightly sloping or rounded, although firm and strongly muscled. The tail is well joined, curved, with long fine hair, and often slightly raised. The skin is thin and covered with fine, short, shiny hair. The usual colors are black, grey, sorrel, and dun. Most have dark skins, and their weight varies from 750 to 950 pounds.

When the Peruvian *Costeña* is resting, his appearance does not attract much attention; but when he is mounted, he shows vitality and style. He is most comfortable to ride, particularly in his principal gaits of *paso llano*, *entre paso*, and *aguilillo*. These have a quality that cannot but interest and arouse the enthusiasm of a horseman, even though he may prefer a larger or different type animal. His easy gaits are especially popular with those who ride for pleasure.

The Peruvian *Chola* is more like the *Costeña* than the *Morochuca*. For the most part the characteristics of the *Chola* are the same as the *Costeña*, with the following outstanding differences: The *Chola* has smaller ears and the forehead sometimes projects a bit around the eyes; the neck is heavier and shorter, and the mane and skin is not quite so fine. The croup is more inclined, and the general conformation is more angular. The *Chola* has some of the same gaits as the *Costeña*, but he is not quite so stylish and does not lift his feet quite so high. He is stronger and of a quieter disposition. He is a particularly good horse in the hills, and if he is not quite so beautiful or stylish as the *Costeña*, his short, vigorous lines reveal his strength and agility. This type is normally found in the higher altitudes, and from this habitat the horse gets one of his names, "*Serrana*." The *Chola* is generally considered to be the better stock horse, and is so used in Peru on the ranches.

The *Morochuca* lives in the highlands of the Andes in altitudes ranging from 8,000 to 13,000 feet, where the pastures are cold and barren. The *Morochuca* is small in stature, short bodied, and big bellied. He has a narrow chest, a thin neck, and a scanty, coarse mane and tail. His bones are coarse and his form angular. His ears are small and hairy, and his hide is thick and covered with long, wooly hair. In pure resistance he may be the superior of other Peruvian horses. He can travel on the average of fifty miles a day at his famous "*hurachano*" pace. His resistance and stamina are almost incredible, and he withstands the vertical sun rays and biting snowstorms in the high Andes with apparently little discomfort. Although he has little shelter winter or summer, and seldom good food, he continues to give sterling service.

The Peruvian horse has changed greatly since the day of the conquest. The *Chola* is most like the horse of the conquest, although even he is changed. The Peruvians believe that their *Costeña* is more beautiful, and their *Chola* and *Morochuca* more useful than the horses of the conquest.

For many years no effort was made to bring back the old-time native horse of Peru. In February, 1913, in the agricultural magazine *La Riqueza Agricultura*, there appeared the first sign of such a movement in an article entitled "*El Caballo Peruviano*," which called attention to how dangerously the horse population had been decreased in the past seventy-five years. The article pointed out two principal reasons for this change: first, the indifference of the agriculturalists; and secondly the ignorance of the great number of people of the worth of horses. The article claimed that the Peruvian *Criollo* was only a pale shadow of what he had been in times past, adding that the natural attraction of things foreign and the indifference to things national had had its effects on the Peruvian horse. The Peruvian *Criollo*, according to this article, is a race well defined, produced by the native environment of Peru and irreplaceable for Peruvian use. As a result, today there is a movement on foot to preserve and encourage the breeding of the native horses of Peru.

During the last twenty years there has been a surprising amount of interest in the Peruvian Paso horse in the United States. Perhaps the gait of the horse has had much to do with his popularity. This gait, distinctive to the Peruvian horse, is commonly called *termino*. It is an exaggerated movement of the front legs with the hoof following an arc instead of a straight line as the horse moves forward. The result is spectacular action of the forelegs, knees, and shoulders of the horse. A larger horse is favored in the United States, and Paso horses of over fifteen hands and of twelve hundred pounds are not uncommon. Generally normal colors are favored.

At present there are about two thousand registered Peruvian Paso horses in the United States and almost two hundred breeders in twenty-six states and Canada. Unfortunately, the three associations that have arisen do not always work together. Puerto Rico also has the National Association of Paso Fino Horses.

Caballo Chileno

The descendants of Andalusian and Castillian horses, when adapted to the new surroundings in Chile, founded a breed which is now called the "*Caballo Chileno*." Although the Chilean horse is not designed for prolonged speed in racing, his environment has strengthened his quarters, lowered his withers, and given him tremendous muscles for his own particular life. His development seems

Rizado, a champion Peruvian Paso gelding in traditional Peruvian gear. Note the *termino* gait. (Photograph by Foucher, Guerneville, California)

to parallel in many respects that of the American Quarter Horse.

It was in the nineteenth century that the *Caballo Chileno* was definitely established as a breed. There were by that time many ranchers, located especially between the Valley of the Bio-Bio and the Copiapo, who were raising these horses. In the area of Aconcagua, Santiago, and Colchagua, the final basis of the best traditional genealogical families arose and have been retained until today. Surprisingly enough, distinctive and homogeneous types seem to have existed even in the nineteenth century. Among them were the *Caballos Cuevanos*, established by the Donihues of Quilamute, and the *Choapinos* of *El Principal de Catemu.* These horses all showed the results of continuous and conscientious selection.

A few of the best breeders of Chilean horses always preserved the purity of their animals from all outside crosses. The imported animals were for the most part of European origin and blood, and were first introduced about the middle of the nineteenth century. The earliest Thoroughbred came from Australia in 1845. Following the Thoroughbred, the coach horse entered from Europe in 1860. The Chilean horse was protected from outside blood during the early years and up until 1800 by Spain's economic policy of controlled trade, and since that time by the genealogical record kept by the National Agricultural Society (*Sociedad Nacional de Agricultura*).

The National Society of Agriculture was among the first organizations to stimulate and encourage the breeders' initiative and desire for a truly Chilean horse. It was mainly responsible for the early establishment of the studbook. The studbook was opened in 1893 and included, as it should, the names of the best animals belonging to the oldest breeders established in the country. Registered as Number 1 was the stallion Bronce, owned by Don Diego Vial Gusmán. Bronce was foaled in Aculeo in approximately 1882 and was bred by José Letelier Sierra. He came from a line of horses that José Letelier Sierra and his brother, Don Wenceslao Letelier, had bred and improved for many years preceding the foaling of Bronce. The Leteliers' holdings were in Vichiculen, province of Aconcagua, and in Aculeo in the province of Santiago. All the ancestors of Bronce are known for at least three generations.

Following Bronce in order in the studbook are the two mares Vinda and Novicia, who were also the property of Diego Vial Gusmán and bred on his ranch, of *Cuevano* stock. Codicia, the sire of Bronce, is also registered as Number 73. Codicia was born in Aculeo in 1876. He, too, was bred by Don José Letelier Sierra. Bronce sprang from a family that is one of the most popular of all Chilean horse strains.

In the first few pages of the Chilean studbook is a list of the principal families of Chilean horses, the work of Don Francisco A. Encina. The Chilean horse families are found for the most part in certain areas, and each family has several important strains. These families, bred and raised for generations and many times interbred, transmit to their descendants a conformation and certain characteristics, even today. In some families, "*Aculeo*" and "*Cuevano*" for example, one can readily see the characteristics of the Castilian horse, while others, such as the "*Principal*" and "*Catemu*" families, show their Barb and Andalusian descent more than their Castilian. This is no doubt due to the blood of some progenitor who was strong in these characteristics, and doubtless increased by the desire of certain breeders who selected and bred for these qualities.

In spite of the enthusiasm of the breeders and the various thousands of horses, there were only 262 animals registered up to and including 1900. This was probably caused by lack of realization of the importance and value of the registry for the Chilean horse. They were then too common. In addition, there were no other competing breeds at that time with which to compare the Chilean horse in order to judge its worth. It is also possible that there was, to some extent, a tendency to encourage registration from certain breeders. It was undoubtedly because of this that we find about 1894 another registry of Chilean horses being formed, under the sponsorship of the Draft Horse Association (*Associación de Criadores de Caballos de Tiro.*)

There were other factors restricting the development of the Chilean horse in the early years. One was the severe agricultural crisis which lasted from 1896 to 1898; another was the sudden wave of popularity for the coach horses; and then, too, there was the decreasing interest in country life. There was one other factor, which was later to appear in North America, which likewise retarded the breeding of the Chilean horse—the inclination of the ranchers for a tall, leggy horse, possessing but little value as a using animal on the ranch. The numbers of horses at the various fairs and shows decreased to such an extent that in 1909 there only was a single animal shown, Huiche by name, owned by Don Diego Vial Gusmán. This great breeder, with a tenacity of purpose and an unshakable belief in the value of the Chilean horse, kept right on breeding until once again the native horses gained recognition.

Finally, in 1910, a group of wealthy ranchers and horse enthusiasts decided to devote themselves to breeding and promoting Chilean horses. As before, the National Agricultural Society was ready and waiting to assist them, and created a special committee and section for the breeders of Chilean horses. This was promptly enlarged to include forty names. The first work of this section was to see that there was only one registry for the Chilean horse and to eliminate and discard the registry maintained by the Draft Horse Association.

The first step taken was to investigate carefully the pedigrees of all registered horses. All the animals whose pedigrees were questionable were eliminated. After this was done, all the remaining animals were personally inspected, and if satisfactory were recorded in the rejuvenated studbook of the Chilean horse. For a number of years thereafter this committee was in charge of rigorously inspecting all horses and pedigrees before they were registered in the studbook. Repeated trips were made up and down the country, inspecting horses and lending enthusiasm and life to the expositions, fairs, and rodeos.

By 1920, one decade later, the number of registered horses had risen to 2,244, and in 1936 to 6,550.

A work equally as important as the reorganization and vitalization of this registry was carried on during this same time by the breeders. They gathered up all the good stallions and mares available. By good fortune there was found in Aculeo the stallion Angamos I, who has had more influence on the breeding of the modern Chilean horse than any other. Also discovered were some excellent mares of the *Principal* family, the best descendants of Guante I, and the stallion Petizo, of the *Cardonalina* family. Don Tobias Labbe found the great stallion Gacho also of the *Cardonalina* family. Additional valuable mares of other origins and also other stallions were found.

During the last twenty years, there have been continuous crosses between the different families which originated both with the registry and before it. The guiding factor in the Chilean crosses has been the effort to breed a horse as close to the "Standard of the Race" as possible. A statue of the ideal Chilean stallion, generally considered practically perfect in every way, was modeled by an outstanding Chilean sculptor, Don Francisco Casas Basterrica. The model was the stallion Ayahas, No. 1046, although the statue corrects a few slight defects which he possessed. The association considers this statue to represent the ideal *Caballo Chileno*, and in truth, he is as good a representative of the breed as one could wish.

Standard and Description

The standard or characteristics of *El Caballo Chileno*, as he is called, are described in the Chilean studbook. They are divided into two sections: first, the general characteristics; and secondly, the breed characteristics. A translation of the standard follows. It is interesting to note that the horse's faults are noted in the official standard and improvements suggested. Most standards admit no faults.

Generally speaking the Chilean horse, with regard to type, is a muscular, strongly built horse, agile and rapid in all of his movements. The measurement of his body [passing over his withers and going under, just behind his forelegs] varies from 63½ inches to 71½ depending on his height, which runs from 12-2 to 14-3 hands.

The specific breed characteristics of the Chilean horse are as follows: Although any color may be found, the horse is generally a solid color with a little white on the head and legs. The hair is usually heavy, with the mane abundant and wavy. Very little hair is found on the fetlocks.

The head is not heavy, is of medium length, with a broad, flat forehead. The profile is straight or slightly convex. The eyes are bright and often covered slightly by the upper eyelid. The ears are small and mobile. The neck is of medium size, wide at the base and firmly inserted into the shoulders. It is slightly curving in its upper line, full and almost rectangular in the lower part, and fine in its union with the head.

The withers are generally too low, short, and hidden in the muscles. This fault should be improved following the pattern of a good saddle horse. The withers should be more prominent, yet without the exaggeration of certain other light breeds. His shoulders are strong and muscular, but at times too short and straight. Length and slope of shoulder are indispensable in any good saddle animal that has agility and movement.

The chest is broad, muscular, and well rounded with a good distance between the shoulder blades. The back is also well muscled, although at times a trifle long and low where it meets the withers. It should always be strong, short, and well rounded, generous at the withers, at the loin, and at the croup. The loin is ample, muscular, and strongly united to the croup (generally in a straight line, although at times slightly curved). The croup is large, full and slightly sloping. The base of the tail is a bit low. The tail should always be carried flowing, or slightly lifted. The trunk is strongly developed, with finely arched ribs, well-rounded belly, and short, full flanks.

The muscular forearm is large, straight, and well tapered, with the elbow independent of the frame. The knee is strong and broad, the cannon is moderately long and relatively slender. The legs should be ample when viewed from the side, with strong, free tendons. With regard to the hind legs the thigh is well muscled, the quarters are large, and the leg well muscled inside and out. It is a characteristic of the race to move with hind legs rather far apart. This quality shows strength and energy but should not be abused or the harmony of movements will be lost.

The fetlocks should be small, round, and with very little hair. The pastern is short and strong, with sufficient slope to give elasticity. The hoof is relatively small and high, with a concave bottom and with the frog not overly developed. His temperament is docile, but energetic notwithstanding.

Emilio Solanet, Creator of Criollos

The *Criollo* horse that is found in Argentina, Uruguay, and southern Brazil does not owe its existence so much to several great horses, such as Godolphin Barb, the Byerley Turk, and the Darley Arabian, or to one great foundation animal such as Justin Morgan or Hambletonian, as to one man, Dr. Emilio Solanet. In other breeds of livestock, when one talks of the beginnings, he speaks of great foundation animals; when discussing the *Criollo* breed of horses, he speaks of Solanet. The reason for this is twofold: first, he was the motivating force in organizing the breeds; and secondly, the *Criollo* had already acquired racial traits when he began his work. Today the name "Solanet" when connected with a horse often means more than breeding or ownership, it means *Criollo* of the accepted "standard."

Solanet's original selection of animals, which has proved so excellent, fixed the ultimate form and conformation of the *Criollo*. The universal acceptance of his original judgment of native horse-

Above: An Argentine *Criollo*. (From a photograph by Thorlichen). Below: A group of *Criollos* on Emilio Solanet's ranch. (From a photograph in the author's possession)

flesh is reflected by the almost unbelievable consistency with which his horses have won championships wherever shown. Solanet is not only a man who knows a good horse when he sees one; he can also breed one, a far more difficult task.

In private life, Emilio Solanet is by choice a professor in the Agricultural and Veterinary College in Buenos Aires. He teaches because he likes to. For many years he had been interested in the native horses of Argentina and collected them on his beautiful ranch, *El Cardal*. He studied their origins both in Europe and in the New World. As early as 1900 he was eagerly devouring all possible information on the native horses of Argentina, their habits, and their history. The material he gathered during these years has become doubly valuable today. His books, photographs, and notes are extremely significant because he not only collected information personally but enlisted the help of his older friends, some of whom had been closely in touch with the native horse during the nineteenth century. He made many trips into the interior of Argentina, searching for horses that were free from all blood except that brought in by the Spaniards.

By 1910 he had found in the vicinity of western Chubut the type of horse for which he had been searching. It was not until ten years later, however, that he was able to bring home successfully a group of these horses. After encountering great difficulties, he gathered approximately two thousand horses, from which he carefully selected fifteen mares which were to form the basis of his brood stock. These he combined with a few other individuals that he had secured from other parts of the country. In his original herd he had carefully selected as physically perfect a group of mares as possible, all typical of the *Criollo* horse. They were of the type that had established itself as a breed, in the loose sense of the word, during the three hundred years the horses had been living in the Western Hemisphere.

The registered *Criollo* horses still show characteristics of the bloodlines which formed the foundation of the horses originally imported by the Spaniards. Oriental influences can be seen in the convex shape of the head of some *Criollos*, while other individuals will show the curved head of the Barb. Solanet personally seems to prefer the latter type, as he believes this animal has a stronger bone structure. However, today, with the majority of the breeders desiring the convex structure, he is no longer stressing the Roman noses or Barb type of head.

As the years have gone by, Solanet has continued breeding *Criollos*, and has seen the movement grow until today not only Argentina but Uruguay and Brazil are all working together breeding and preserving *Criollo* horses. The breeder of the Chilean horse is also co-operating with his *Criollo* horse, which he calls, as has been pointed out in a previous chapter, a *Chileno*.

Perhaps Emilio Solanet's greatest opportunity came in 1925 when his young Swiss school-teaching friend, Aime Tschiffely, decided he wanted to see the rest of the Americas. It was Tschiffely's plan to ride north through South, Central, and North America from Buenos Aires to Washington. Tschiffely approached Solanet for assistance in obtaining the right horses. He could not have spoken to a better man. Tschiffely, with Solanet's help, selected two *Criollos*, and the ride was a success. In fact, when the writer was at *El Cardal*, Solanet's ranch, in 1943, Mancha and Gato, who made the long and arduous trip, were still alive and frisky, both over thirty years of age.

The organization developed by Emilio Solanet and other *Criollo* addicts has not been a success solely because it recaptured a type that was fast becoming nonexistent. It has been a success because it is promoting a breed which many believe has never had an equal for working livestock and for all-purpose saddle work. The descendants of the Spanish horses in both North and South America became superb cow horses and enduring saddle animals.

Although Emilio Solanet rightfully is accorded the greatest

honor for starting the *Criollo* movement, there are additional men in Argentina who have been equally enthusiastic and great breeders, such men as Roberto Dowdall, both father and son, the Ballesters, the Lastras, and others who will never be forgotten in the annals of *Criollo* horse lore. Nor will the author forget the pleasant days he spent with them, drinking maté and talking horse.

☙ *Criollo Characteristics*

As a result of the natural selection which took place on the pampas after the advent of the horse in the sixteenth century, the *Criollo* or native horse has a marked resistance against weather and privation. He may have lost some height, weight, and beauty, but not ability and endurance.

With selection and supplementary feeding, there are now arising some excellent examples of *Criollos*, who possess increased style and refinement. More of this type are being registered in the *Criollo* studbook each day.

The *Criolla* Horse Association was established on June 16, 1923. There was a large stock pile of material already available on the *Criollo*, thanks to men like Emilio Solanet. This material covered their introduction on the pampas, the haphazard selection on the *estancias*, or ranches, and the scientific selection which had taken place on various stud farms commencing in the twentieth century.

Justo P. Sáenz (*hijo*), Enrique Lynch Arribalzaga, Emilio Solanet, Deciderio Davel, Roberto C. Dowdall, Angel Cabrera and many other writers have interested themselves in, and written about, *Criollos*. Much of the present wide knowledge of *Criollo* habits, customs, characteristics, and value, is due to the efforts of these men.

The *Criollo* Breeders Association is under the guidance of the Rural Association of Argentina, which is a national organization responsible to the government.

Above: Mancha and Gato, Aime Tschiffely's two *Criollo* horses, photographed with the author (left) and Tschiffely at El Cardal, Argentina. Below: Breaking a horse in Corrientes, Argentina. (From photographs in the author's possession)

The coat colors preferred are those which are generally prefixed in the United States with the word "line-backed," whether they are almost black, tan, yellow, or red (grullo, dun, palomino, or sorrel.) The head should be short and pyramid shaped, the neck well joined, the withers muscular, the back straight, the loin short, the croup large, the tail firmly fastened, the chest ample, the belly cylindrical, the shoulders large, the flanks short and full, the legs true, the forearm muscular and long, and the cannons short.

The *Criollo* may appear in almost any coat color, although since 1920 there seems to be a tendency to favor *gateados*. The word *gateado* is difficult to translate. The Spaniards have a much more adequate color terminology for horses than the English. There are fourteen general color names, similar to our bay, sorrel, grey, roan, brown, and black. There are sixty-seven common adjectives to supplement these basic fourteen color terms, and an additional twenty-seven terms for colorations located on any special part of the body. There are also twenty-three terms for colorations peculiar to the head, and fourteen for those peculiar to the feet and legs. All in all, this gives adequate descriptive words for all colorations but makes accurate translation next to impossible. To return to *gateado*, the best translation is "line backed with black points." However, all are not duns. Celebrated *Criollo* animals, moreover, have existed in all colors, although *tostada* (dark sorrel), *moro* (blue roan), and *gateado* are favorites.

According to the Argentines, the *gateado* is particularly favored in that his coat color gives him a natural camouflage. Many armed forces have adopted a similar color for the uniforms of their soldiers—in order that they be less visible at a distance. Some of the original horses that the Spaniards brought to the New World were of this color. However, while the horses roamed the pampas in a wild state, the *gateados* seem to have increased all out of proportion to their original numbers. Perhaps this was a tendency of nature to help protect them from easy visibility and in this manner give them another safety factor against animals such as the puma, the guanaco, and the jaguar.

The *Criollo* is a small horse but very active. His maximum height is considered to be 1 meter, 52 centimeters (about 14 hands, 3 inches) and the minimum 1 meter, 40 centimeters (about 13 hands, 2 inches). Greater height is held to be a symptom of lack of purity. Those that are smaller in height, for example 1.38 to 1.40 (under 13 hands, 2 inches) may still be good *Criollo* horses, but they are justly criticized as lacking height, and commonly referred to as "*petizos*."

The perimeter of the thorax (or chest measurement) varies from 1 meter, 70 centimeters to 1 meter, 86 centimeters (from 66 inches to 72 inches.) This measurement is considered very important. Many people favor a taller horse than a *Criollo*. The Gauchos of South America claim an increase in height means a diminished strength. They say, and with some justice, "*Despreciar el grande y ensillar el pequeño*" ("Reject the big and saddle the small").

The same theory has been advanced by European horse specialists. Professors Baron and Crevat of Europe, with their widely accepted formulas, are a case in point. They claim that their experiments show that if a horse is 1 meter, 52 centimeters (14 hands, 3 inches) in altitude and 1 meter, 86 centimeters (72 inches) in the thorax, he can carry 127 kilos (about 280 pounds) easily. Another horse of greater height—say, 1 meter, 60 centimeters (about 15 hands, 3 inches) but equal in the thorax, for example—could carry only 121 kilos (about 266 pounds). The taller, leggier animals are more stylish for parades but less capable of work.

There are horse fanciers who believe this inconvenience disappears where the thorax is increased proportionately. However, Dr. P. Dechambre proved experimentally the fallacy of this conviction. When the thorax is increased, the horse becomes too heavy. When a horse exceeds 400 or 500 kilos (900 to 1,100 pounds), he loses some of his value as a saddle animal. With the extra weight, the animal loses not only agility but speed, for it is too much work to carry his own weight around very fast. His tendons and legs break down, and he becomes coarse and lazy, eats a great quantity of food, and prefers to walk. This would indicate that the Gauchos may know what they are talking about—that the horse 1 meter, 45 centimeters (about 14 hands, 1 inch), with a corresponding thorax, is the best for the pampas cattle work. In North America, the cattlemen and cowboys prefer a horse that runs between 14 hands, 2 inches and 15 hands, 1 inch.

Each year breeders from several South American nations gather at the Reunion Interamericana de Criadores de Caballos Criollos. The idea for an international get-together was first proposed in 1942 by Guillermo Echenique of Rio Grande do Sul, Brazil. I had the good fortune to be living in southern Brazil at that time and, being a friend of Echenique, became involved in the movement. Key figures who were present and signed the original papers were, besides Echenique, Roberto Dowdall, of Buenos Aires, Argentina; Juan de Arteaga, of Montevideo, Brazil; and Miguel Lettelier, of Santiago, Chile. All were *Criollo* breeders and outstanding citizens of their countries.

On April 21, 1973, I received the following interesting letter from my old friend, Roberto Dowdall concerning recent developments. What follows is a free translation of his letter. At the fourth meeting of the *Criollo* association, held in Buenos Aires, a common standard for *Criollo* horses was drawn up by the participating members and taken back to their respective countries for approval. It was officially adopted at the 1959 meeting of the breeders. Since that date all registered horses of one country are accepted by all the other nations.

Another advance made by the Inter-American Association was the adoption of uniform rules to cover all official shows, contests, and sales. In 1972 the loose ends were tied together in a more formal organization named the Federación de Asociaciones de

Criadores de Criollos. At present it includes only about one half of South America (Brazil, Argentina, Chile, Uruguay). Other countries are planning to enter soon, particularly Venezuela and Paraguay.

One of the outstanding activities of the South American *Criollo* breeders' association is their annual endurance contest. This remarkable test of equine endurance and hardihood was started in 1954. The distance the horses travel is about 465 miles. Each horse must carry for the entire distance 265 pounds of man and tack. The ride is spaced over fifteen days, and no supplementary feeding is allowed. All the horses are allowed to eat is grass from the countryside.

Before the ride begins, all the horses are assembled at a chosen ranch. The mares are turned out into a common pasture and the stallions assigned identical individual paddocks. The horses are then held for two weeks before the test begins, to be sure that all receive the same care and treatment in advance. They are under the constant supervision of ride officials during the two weeks, and owners and riders are not allowed in the area.

The ride itself consists of fourteen daily stints. Each day's ride covers from 20 to 40 miles, and is generally completed in the morning or by the early afternoon. The route to be followed (different each day) is carefully marked. As each horse and rider finishes, they are separated. For the mares this means the common pasture; for the stallion, his individual paddock; for the rider, the housing area.

Of the fourteen daily stints the first three are the shortest to help the horses who have been idle for two weeks. The last eleven stages are the serious ones, in that all must return to their pasture or paddock by a certain time. If they come in earlier than other horses, this is not an advantage. If they are behind the scheduled time, it does count against them, and that is why these are important days. The last five days are the critical ones. Each rider makes the fastest possible time then, because the total time for every ride is logged, and the horse that has completed the ride in the shortest over-all time is the winner.

After each day's ride, as the horses return to the ranch, they are met by a veterinarian, who examines them and takes their temperature and their pulse. The examination is repeated in half an hour to be sure that each horse recuperates normally. The veterinarians do not hesitate to scratch a horse if any signs point to overfatigue or if any other problem exists. The only attention a horse may receive when he comes in from a day's ride is to be washed down. The rider is allowed to wash the back with soap. Any other care or assistance given to a horse, such as massages, injections, or medicine, will eliminate him from further competition.

While everyone enjoys the competition, it is not held just for sport. It is a test of the *Criollo's* endurance, hardihood, and ability to work under range conditions. It is used as a guide for selection of superior individuals. The *Criollo* breeders brag about the toughness and the endurance of their breed, and, with this competitive endurance ride, they prove it. They also feel that it tends to keep the rugged self-sufficiency of the *Criollo* from deteriorating.

The value breeders place on the successful completion of the endurance ride is reflected by the action of the better breeders. They try to have as many successful animals in their breeding program as possible. Some now have all of their herds of brood mares and stallions composed of successful endurance horses.

Criollo breeders have other means of evaluating their animals. There are hardly any fairs or expositions, either local or national, that do not have classes for *Criollo* horses. These include running, cutting, roping, and cattle-tailing events. Each is rather different from those held in the United States but require equally well-trained horses and skillful riders.

In June, 1973, was celebrated the fiftieth anniversary of the founding of the Argentina *Criollo* horse association. Of all the men who came together to establish the new breed, four were still living, Emilio Solanet, Filipe Armadeo Lastra, Paul Elicagaray, and Dowdall.

Criollo characteristics are alike in all of the South American countries. The differences found are due more to variation of environment than to a difference in standard. For example, on the plains and sandy lowlands, the feet are larger than in the hilly country. Likewise, the Argentine *Criollo* runs about two inches taller than the *Criollo* of Rio Grande do Sul, Brazil.

Brazilian Breeds

Most of the Brazilian *Crioulos* (Portuguese for *Criollo*) are found in the southern state of Rio Grande do Sul. This state is primarily a cattle-producing area which borders both Uruguay and Argentina. The cowboys in Rio Grande do Sul are called, as are their brothers in Uruguay and Argentina, Gauchos.

Although the *Crioulo* horse had been used throughout the state of Rio Grande do Sul for several hundred years for transportation, for ranch work, and for warfare, he was ignored as a breed. It was not until the late eighteen hundreds that some rudimentary selection appears to have been made. Most common throughout the nineteenth century was the crossing of *Crioulo* mares and English Thoroughbreds. Certain other European and Asiatic breeds also were imported at this time, such as the Arab, the Hackney, the Andalusian, the Norman, and the Orloff.

The Rio Grande do Sul Gaucho loved a race, as do all horsemen, and this enthusiasm was responsible in part for the large amount of Thoroughbred blood found in many of the Rio Grande do Sul horses. The idea of improving a strain of livestock by crossing it with some other breed which was strong in desirable characteristics was so popular at that time that the original *Crioulo* type became relatively scarce by the nineteen twenties. By this time it was also obvious to the breeders that their new cross-bred horse had less resistance, could not work as hard, and had to be pampered with special food. Their need was obvious, and a popular demand arose for the old *Crioulo* horse of their youth.

The beginning of this trend toward the old-type *Crioulo* came at the turn of the century, and since 1916 it has been encouraged not only in expositions but also by the government. The native *Crioulo* horse also received favorable publicity in the book by Delfino Riet published in 1919.

The *Coudelaría Nacional de Saycan* (a Rio Grande do Sul remount station) was beginning to improve the *Crioulo* horse in 1927. According to a former director, a *Crioulo* stallion from the *Estancia* Santa Ambrozina of Rosario, Rio Grande do Sul, was used on some *Crioulo* mares which belonged to the remount depot. Later, however, the mares were served by a purebred Arab. It was thought at the time that the *Crioulo* was a degenerated Arab and that, by using an Arab, the old-type *Crioulo* could be obtained quickly. February 28, 1932, marked the beginning of the last period in the development of the modern *Crioulo* of Rio Grande do Sul. At that time there was organized by Guilherme Echenique Filho in Bagé (the Fort Worth of Rio Grande do Sul) an Association of *Crioulo* Horse Breeders (*Associação de Criadores de Cavalos Crioulos*), whose duty it was to select and improve whatever true *Crioulos* could still be found in the state. The Bagé movement reflected a trend toward the old-type *Crioulo* or native American cow horse that was already active in Chile, Argentina, and Uruguay, and which some ten years later was to cause the breeders of North America to create similar organizations.

Rio Grande do Sul had a difficult job: namely, the elimination of the foreign blood and the establishment of a horse that would breed true. Some breeders, like Guilherme Echenique Filho of Pelotas (who was the first president of the *Crioulo* Breeders Association), imported basic stock from Emilio Solanet's ranch in

Argentina. Echenique was, and is today, one of the outstanding, South American agriculturalists and a true horseman and Gaucho.

According to the standards set up by the *Crioulo* Breeders Association, the ideal *Crioulo* has the following characteristics:

The head should be short and in the form of a pyramid, wide at the base and fine at the point, with the jawbones well developed. The cranium is roomy and the face short, the forehead large and well developed. The profile should be straight, the ears small, active, and set well apart at the base. The eyes are large, wide set, and reflect gentle, kind intelligence.

The neck is joined to the head by a clean throatlatch. It is slightly convex in its upper line, with a thick and abundant mane. The neck is almost straight in its lower line and is large and muscular where it joints the body. It is rather more short than long.

The withers are muscular, not particularly prominent, but strong. The back is straight, short, wide, and well joined to the withers and loin, showing the capacity to support and carry weight. The loin is short, broad, muscular, and well joined to the back.

The croup is of medium length, muscular, strong, well developed, and semi-oblique. The tail is thick and short, with the root well placed.

The breast is big, wide, deep, and strongly muscled. The sides are deep and arched, and the chest well developed, a large circumference typical and desirable.

The belly is cylindrical in shape, voluminous when the animal is full and reduced when the food is digested. It is only slightly convex and is joined smoothly with the thorax and the flank.

The flank is small, short, full in relation to the shortness of the loin, oblique, and separated from the ribs.

The shoulders are satisfactory and in proportion to the head and neck. They are sloping, well developed, strong, and well separated. The arms and elbows are strongly developed, the forearm is muscular, long, large, strong, and plumb. The knees and cannon are short, wide, ample, and thick, with strong, clean detached tendons.

The ankles are dry, round, strong, and clean. The pasterns are strong, short, wide, thick, and clean, with average incline. The hoofs are in proportion to the body—hard, compact, solid, well supported, and preferably black. The legs and thighs are strong, well developed, firm, elastic, and muscular. The average height is about fourteen hands. The horses seldom vary more than two inches one way or the other.

The weight varies from 900 to 1,000 pounds, and the average *Crioulo* can easily carry 250 to 280 pounds on his back.

The coat colors preferred are dun, sorrel, red roan, bay, and blue roan. Brown, grey, and black are becoming more common. At present, most breeders do not prefer pinto, either *tobiano* or *overo*, since horses of these colors are hard to sell, but the colors are considered good old-time Crioulo colors.

The Brazilian *Crioulo*, like his brothers in the other South American countries, is basically a small, although stout, saddle horse. In an official publication of the *Associação Brasileira de Criadores de Cavalos Crioulos*, entitled *Normas para o julgamento de Confirmação de Cavalos Crioulos* (Standards for Judging Conformation of the *Crioulo*), published by the breed association in January, 1972, the measurements of the Brazilian *Crioulo* are outlined. The height, circumference of the cannon bone, and the diameter of the thorax are as follows:

Height of stallions, without shoes, minimum 1.40 and maximum 1.50 meters (13-3 to 14-3 hands). Height of mares without shoes, minimum 1.39 and maximum 1.50 meters (13-2 to 14-3 hands). The cannon bone of the stallion has a minimum circumference of 0.18 meters (7 inches), and the mare 0.17 meters (6 7/10 inches). The thorax of the stallion and mare is a minimum of 1.68 meters (66 7/10 inches).

As of December, 1972, registrations in the Brazilian *Crioulo* studbooks were as follows: in the temporary studbook, 6,722 stallions, and 11,916 mares, a total of 18,638; in the permanent studbook, 117 stallions and 183 mares, a total of 300. There were 467 active regular members of the association, and 5 honorary members.

The registry is not open in the sense that all descendants are automatically eligible. All the stallions and mares receive their certificates only after approval by the association. This approval, called confirmation, is given by a special inspector when the horse has reached two years of age. After five generations have been inspected and confirmed, the fifth-generation horses receive permanent registration, and their offspring may be registered without inspection.

Originally the *Crioulo* was found almost exclusively in Rio Grande do Sul. Today, however, they are being registered in the states of São Paulo, Santa Catarina, Paraná, Rio de Janeiro, and Mato Grosso.

The temperament of the *Crioulo* is kind, active, intelligent, and courageous.

Besides the *Crioulo* of Rio Grande do Sul, there are two other breeds which have been developed from the native Brazilian animals. These are the *Mangalarga* of São Paulo and the *Campolino* of Minas Gerais. The *Mangalarga* was started at the time of Emperor Peter II of Brazil, who imported an Alter stallion (Portuguese equivalent of the Andalusian) by the name of "Sublime." By 1857 the breed was rather well established. It is quite different from the *Crioulo* and averages fourteen hands, three and one-half inches in height. Most of the added height is gained in the cannons and pasterns. The Mangalarga's withers are higher than the *Crioulo's*, and his tail is set higher. His neck is more refined, and he is not so deep through the chest. His best gait is a trot. One of the greatest of all *Mangalarga* breeders was *Coronel* Francisco Orlando Diniz Junqueira.

The other Brazilian breed, the *Campolino*, is a native of the state of Minas Gerais. He is a heavier horse, somewhat more like the

Crioulo than the *Mangalarga*. The *Campolino* has shorter cannons and pasterns than the *Mangalarga* and a deeper chest. He is suitable for light draft work as well as for the saddle. The name *Campolino* comes from the originator and principal breeder who lived in the middle of the past century, Cassiano Campolino by name. His greatest stallion was called "Monarcha" and was half Andalusian. In the state of Mato Grosso there is also a saddle horse known as the *Pantanero*. He is found in the southern part of the country, near the border of Paraguay. Tradition, national pride, and utility combine to keep these Brazilian breeds active.

Epilogue

14. Historical Résumé

Sum and Substance

THE HORSE IN LATIN AMERICA

Horses are found in a domesticated state all over the world, in every climate and on every continent. It is fortunate that man's most useful friend is thus a cosmopolite. As wild animals, however, horses are today limited to the Old World. In America there have been no wild horses in the true sense since the Glacial epoch. The mustangs belong to a later date and are semiferal animals, descendants of domesticated horses imported from Europe.

The Spaniards first brought horses to America. They had the finest animals in Europe. Their short backs joined well-muscled shoulders and croups, and their legs, not too long and firmly jointed, made them sure on their feet. Once adapted to the American climate, the horse lost weight and beauty but was compensated by increased stamina. Some almost unbelievable feats of endurance have been recorded.

The horses were first taken to the West Indies by Columbus at the close of the fifteenth century and were rapidly acclimated. Within thirty years the island animals formed the chief supply for the mainland expeditions. From the islands they were taken to the Isthmus in 1514 and to Mexico by Cortés in 1519. Once the conquest of Mexico was over, they were raised extensively in the pueblos, haciendas, and missions of the mainland settlements. Bayamo in Cuba and Tlaltizapán in Oaxaca became the great horse marts, while Gracias a Dios in Neuva Valladolid and Nextipaca in Nicaragua monopolized the mule market for the immense trans-Isthmian trade.

Horses were taken to Peru by Pizarro in 1532 and from there into Bolivia by Gonzalo Pizarro in 1535 and to Chile by Valdivia in 1541. They were taken into Venezuela by Federmann and into Colombia by Quesada about the same time (1538). The first horses came to Argentina in 1535 with Mendoza and thereafter entered from three directions, from the east by the way of La Plata, from the north by way of Bolivia and Paraguay, and from the west by way of Chile. The first horses into Paraguay came north from Argentina in 1536, and others came with Álvar Núñez in 1542, entering by way of Brazil. Still others came in from Bolivia. Martim Affonso brought horses to Brazil in 1530, although Álvar Núñez and the Jesuits from Paraguay were also responsible for the horses in southern Brazil and Uruguay. As early as 1580 travelers began reporting the immense herds of wild mustangs to be found in this territory. In South America, tradition has Pedro de Mendoża responsible for the spread of the mustangs throughout the pampas land, but it is more likely that horses from Chile and Paraguay were responsible.

From Spain, the islands, and Mexico, the horse was taken by the Spaniard into North America in four salients, namely Florida, Texas, New Mexico, and California. While Jamestown was being settled by the English on the Atlantic coast, Oñate was establishing ranches in New Mexico, principally around Santa Fé. The common belief that the mustang of the Western plains arose as the result of horses lost in the early expeditions of Coronado and De Soto is incorrect.

Just as the horse of western North America came from Mexico, so the foundation stock of the Atlantic coast of North America came from the Spanish islands to Florida. By 1650 this district of Guale (northern Florida and southern Georgia) had seventy-nine missions, eight large towns, and two royal haciendas. From these centers the horses spread to the Indians—not from remnants of De Soto's horses as has been commonly supposed. The continued Indian revolts and English depredations at the close of the seventeenth century tended only to spread the horse north into the English colonies.

Thus, the first modern horse of America was bred in the Spanish islands. From the islands he went north by way of Mexico and Florida, and south by way of the Isthmus, Peru, and Chile. Indeed, most of the horses used for the conquest and development of the Americas before the end of the eighteenth century were obtained from colonial Spanish ranches, most of them on the islands of the Caribbean, rather than from Europe. Until the nineteenth century, these Spanish horses presented the main, and in many respects, the best available horseflesh.

THE HORSE IN THE NON-IBERIAN COLONIES

The English, Dutch, and French brought very few horses from Europe before the nineteenth century. When Captain Smith left Jamestown, the colonists ate their mounts. Virginia received a few horses in 1613 from Argall's French venture, although most of these were returned to England. The Canadian Indians got their first glimpse of the "moose of France," as they called the horse, when a shipload came from Le Harve in 1665. Except for the horse sent to Montmagny in 1647, these were the first horses in Canada. Their number increased thereafter, according to the Jesuit relation of 1667. The French did not import many horses because they lacked year-round feed and used the canoe for inland transportation. Wood, Sandys, and Gookin, the first importers to Virginia, brought Irish horses about 1620, but the animals were still scarce in 1640. Horses and cattle were imported into Massachusetts in 1629, and in 1638 the Swedes and Finns brought horses to Delaware. In the southern border warfare at the close of the century, the English obtained Spanish stock. On the outskirts of Virginia they multiplied in a semiferal state—like the mustangs of the Southwest—until they were a menace to the crops and were hunted for sport. These horses were generally small, but new blood was obtained from the Spanish stock west of the Mississippi. They were crossed on the few imported English running horses and developed the Narragansett pacer and the Colonial quarter-mile horse, the progenitor of the modern Quarter Horse.

The Dutch obtained many of their original horses from Curação, although two Dutch schooners brought twenty-seven mares from Flanders in one trip in 1660. The Dutch herding policy, like that of the southern English colonies, allowed the stock complete freedom. The Dutch did not employ town herders for the village commons as did the New Englanders. The sugar mills of the West Indies exhausted the island herds, and horses were exported from the English colonies as early as 1656 by Coddington, but not until the eighteenth century did this West Indian trade reach its peak. Some of the New Englanders, seeing the possibilities of this "jockeyship" trade, began raising horses. Hull stocked Boston Neck for the trade in 1685.

New England, with its communal grazing and constant herding, had enough stock for its use almost from the start, the prices of horses dropping from an average of £34 in 1635, to around £2 in 1800. The drop in price was proportionate throughout the colonies. It was due not only to the increase in supply but also to the inferior quality of most of the stock, which had degenerated because stallions of small stature and with bad points were allowed to run with the herds. The better horses were kept for saddle and coach. This helps to explain why the more wealthy men continually imported small numbers of good animals for personal use. Patrick Henry sent to the Pawnee country for the "best and most pure Spanish breed," as he called the horses, many of which did indeed bear Spanish brands.

Loyd Jinkens training a cutting horse. The horse is turning abruptly with a loose rein and a closed mouth, as a good cow horse should. (Courtesy *The Cattleman*)

With the Anglo-American penetration into the Great Plains of North America, we find the colonial English and French leaving their previously essential canoes and axes at the Mississippi River and adopting the horse and Plains culture in place of woodland ways. The Great Plains provided an ideal environment for livestock, and the American cowman replaced the Indian and the Mexican inhabitants. In South America, the horse was from the first utilized in the basic economy of that continent, and no civilization similar to that of seventeenth- and eighteenth-century English woodland colonies arose.

The influence of the horse imported from Europe, particularly of the Thoroughbred from England, was not felt until the nineteenth and twentieth centuries. Then this blood was mixed with that of the native horses. The horse's place in the development of the hemisphere has thus been unique in that from the time of his arrival he has been an integral and irreplaceable adjunct to man's conquest and development of America.

Appendices, Bibliographical Sketch, and Index

Appendices

◆ *Naming the Horse*

It is one of the traditions of the Western horseman to love a colorful horse, whether that color lie in his coat or in his actions, and nowhere will you find a group of men more conscientiously living up to their heritage. There is an old English saying to the effect that no good horse can be a bad color regardless of his coat. But the *conquistadores* must have thought that no good color can be on a bad horse, if we believe some of their more enthusiastic statements.

Don Bernardo de Vargas Machuca is one of the best early Spanish authorities on horses and horsemanship. He wrote in 1600 right after the conquest. "The bay," he says, "is a natural and perfect color for a horse." In spite of this seemingly overbold statement, his sentiments coincide with those of most writers of that period, not to mention several moderns. Not at all satisfied with this generality, he plunges boldly on with statements that would arouse an argument in the most timid horseman. "A few white hairs above the tail are a sure sign the horse will be swift and strong, and have a good mouth [*ser de buena rienda*]. . . . *El Ruzio Rodado* [dapple grey] is very handsome, and most of these horses turn out fast." As Graham so aptly puts it, "It is a well attested fact that some horses are swifter than others." Vargas adds, "The cream colored horse with black points is handsome but seldom fast. . . . The dark chestnut is swift and strong but choleric and has a bad mouth." Whether the weak mouth is due to the "choler" he does not say, but might it not possibly be the breaker's fault?

"The black is handsome and fast, he has a bad mouth, is choleric, and short sighted, jumpy, and treacherous." Many blacks are high spirited, but whether they are always treacherous and short sighted is another matter.

"Pintos are showy," Don Bernardo says, "but subject to disease and always have bad mouths." After some little time considering white horses, he states merely that they are handsome and swift with a good mouth, but the white hairs soil one's cloak. His writings are full of curious superstitions and sayings, but he wisely finishes, "*Pero en esto de colores, camine cada uno a su gusto*" ("In regard to color, each should suit himself").

A favorite Spanish custom was to name the horse according to the impression received on first seeing the animal. Thus we find *Labuno*, meaning "wolf-colored," *Gateado*, "cat-colored," and *Pardusco*, "mouse-colored." Cortés called his horse, simply enough, *Morzillo*, because he was black. Sandoval's horse, considered by Bernal Díaz del Castillo as the best animal in either the Old or the New World, was called *Motilla*, meaning a tuft or a crop, although we are not told where it was located. De Soto's horse was *Azeitunero*, after the name of a prior owner. Velásquez's horse was *La Rabona* because of her extremely short tail. Bartolomé García had a black horse he called *Arriero*; he must have either looked like a mule or worked with a pack train. The list might be extended indefinitely. The *conquistadores* favored a descriptive name, and Westerners today retain this custom.

On Mexican ranches, similar titles are given to the horses: *Venado*, or "deer-colored"; *Zorillo*, generally a dark horse with white points, since the name means "polecat"; *Cuervo*, "black as a crow"; *Pico Blanco*, "white muzzle"; *Nopalero*, "good in prickly pear"; *Pelado*, "hairless"; and so on.

The Anglo-Americans were liable to call their horses by almost any name. If the animal were slow, he might be called Possum or Molasses; if he did not lift his feet well, or was a pacer, they might call him Sand Sifted or Trail Cleaner. If they bought the horse, they would often name him after the seller, as John Brown or Jim Lowe. It was also common to call horses by their markings, as: Star, or White Foot, or Zebra Dun, or Strawberry Roan, or Sorrel Stud. Outlaw horses always had most picturesque names, such as Danger, Dynamite, Bad Eggs, Widow Maker, and Sudden Death. A fast horse might be called Bumble Bee, Wind Splitter, Gander, or Daisy Clipper. If the horse were long with short legs, he might be called Dachshund or Gopher; if he held his tail high, Scorpion; if he shyed, Spooky. Daisy, Nellie Gray, Sweet Alice, and Fay are all self-explanatory.

Quarter Horse nomenclature seems especially picturesque and even mellifluent. For example, Little Joe, Ace of Hearts, Coal Oil Johnnie, Cold Deck, Steel Dust, White Lightning, Fireball, Little Judge, Grey Alice, Hay Seed, Little Nip, Sleepy Dick, Scarface Charley, Billie Sunday, Carrie Nations, Ram Cat, Cut Butt, Rocky Mountain Tom, Cherokee Maid, and Chicasha Bob are just a few of the most prominent appellations.

As names of horses follow the whim of the owner, they take every form, although the originality expressed in the names given Western horses make them most interesting.

◆§ *Evolution of the Western Saddle*

Before the coming of the Moors, Spain utilized the saddle that was common throughout Europe; it was the saddle used in the *a la brida* style of horsemanship of the armored knight. One of its principal characteristics was the long stirrups. There were three main variations in the *a la brida* saddle. The first was the *silla de croata*, used for everyday riding with very light armor, and in some cavalry units it was employed primarily for scouting. The second was the *silla de estradiota*. With this saddle not only was the rider more heavily armed, but the horse also wore some armor. The front and rear portions of the saddle were of metal and partially enclosed the thighs of the rider, for, since fighting was largely a matter of individual combat, metal provided the protection necessary to keep the rider from being knocked from the horse. The third type was the *silla de armas*, used by those knights who were clad from head to foot in steel. The horse was also well covered, the result being a walking fortress, or forerunner of the modern tank. There was one difficulty in this type, however: to mount or dismount, the rider had to have assistance.

Since the horse being ridden *a la brida*—that is, with long stirrups—as well as his rider, was covered with armor, it meant that the horse must be large and powerful at the expense of speed and agility. He needed to be large in order to carry the heavy saddle and armor with which he was burdened.

The Moors who invaded Spain in the eighth century used lighter horses and equipment. They rode *a la jineta*, a style characterized mainly by the bent knees made necessary because of the short stirrups.

There were three principal types of the *sillas de jineta* used by the Spaniards after the type had been adopted from the Moorish invaders. The most magnificent of these was the *silla de jineta para fiesta*. This great predecessor of the famous silver-mounted Western saddle was caparisoned with velvet and bore a delicate border of gold and silver. The stirrups were hardly less elegant, for they were covered with engraving and highly-embossed relief work. The second style was the *silla de jineta para caballeros*. This saddle had a black cover and bridle, varnished stirrups, and cantle cover and reins made of special Barbary leather. The bridle, martingale, and tie rope were made of soft and velvety Moroccan leather. The bit was of highly polished metal inlaid with gold. The last type was the *silla de campo*, or work saddle. The trappings on this saddle were not so rich but more sturdy with less engraving and more sedate in coloring. It was this that was generally used in combat.

The first Spanish American ranchers on the islands of Cuba and Hispaniola became immensely wealthy. Buying everything their fancy dictated, they imported the finest horseflesh from Europe and created a magnificent predecessor to the Mexican saddle, studded with silver and jewels and inlaid with precious metals. However, the saddle makers on the islands were important only for a short time, and they made no outstanding original contribution.

After the occupation of the islands, the Spaniards moved on to the mainland. The saddle developed in Mexico is the direct forefather of the American stock saddle. Some of its features are considered superior by our foremost riders, which explains why American saddles today are so often seen with one or more typical Mexican elements, such as a large flat horn or lower cantle. This saddle not only is still practical, but is used throughout Western America, with certain local modifications.

The Spanish conquerors did not want the Indians to have horses because they felt that they would lose their principal weapon as well as gain formidable enemies. Despite the severe regulations issued to enforce their wishes in this respect, many natives obtained mounts and became expert horsemen. How this occurred is well illustrated by the example of Fray Pedro Barrientos, who, soon after helping pass a law which was to prevent natives from riding, went to his estate to watch his native *vaqueros* give a show of horsemanship with commendable agility. It was in just such ways that the Indians became accustomed to the use of Spanish saddlery, and hence were not satisfied without them.

The law forbidding natives to ride, like so many laws, soon deteriorated, until it merely forbade their using saddles. This worked a hardship on the native Mexicans, who had to ride without benefit of saddle. However, necessity is the mother of invention. When the renowned *talabatero* or saddle maker, Alonso Martínez arrived in Mexico City in the early days, he hung the age-old sign of his profession, a *fuste*, or saddletree, in front of his shop. One night it disappeared. Within twenty-four hours it was back. The natives, recognizing it, took it to use as a model. From this original copy (at least according to the story) all future Mexican saddles were constructed. And, if this is true, from this same hurriedly imitated tree arose the modern stock saddle.

Before the close of the sixteenth century, a viceroy gave privileges to certain native caciques, or chiefs, which permitted them to ride horses. There was, however, a stipulation in the ordinance that they must use the native Mexican saddles. Thus, before the start of the seventeenth century, the difference in the two saddles was recognized.

The riches that were in Mexico, the abundant gold and silver that were found, combined with the Spaniard's very human desire to flaunt their new-gained wealth, gave rise to another variation of the Spanish saddle. This variation was still basically an *a la jineta* saddle, but profusely decorated with trimmings of gold, silver, and silks in beautiful combinations. If any one color predominated, it was blue.

It was not long until the style of riding had to be somewhat modified. As the horses thrived in the new land, growing numerous and wild, the people found difficulty in mounting a half-broken animal with such short stirrups. As there was continuous riding day after day watching the herds, this cramped method proved rather uncomfortable. Therefore, the stirrups were lengthened. The old method of tying the *lazo* to the horse's tail was awkward, too; therefore, it was not long until the pommel was capped to hold the rope. Thus the modern stock saddle had achieved its basic form by the end of the seventeenth century.

When independence from Spain was achieved, there came with it an awakening of a new sense of nationalism. There also arose to power the opulent and influential *campesanos*, or country gentlemen, who discarded the old *a la jineta* saddle and patriotically adopted the native saddle. They retained the custom of adorning saddles with gold, silver, and silk by simply transferring the decoration to the native saddle. This was the greatest change which had come over the saddle since its origin with the Arabians centuries earlier. There had been a continual evolution in workmanship until that time, contributed to heavily by native Aztec craftsmen, and after the adoption, there was increased refinement of decoration and standardization of trappings. With the close of the eighteenth century, the characteristics of the Mexican saddle, as distinct from its predecessor, the *a la jineta* saddle, apparently became fixed.

Some horsemen consider the Mexican rig the greatest roping saddle ever devised. Chief among the characteristics adopted by the modern cowboy are the low cantle, allowing the rider to move away from the awkward throw; the total absence of any swell on the forebow, thus eliminating any chance of the rope's catching under the bow or of slowing up the dismounting; and, lastly, the horn, which fits snugly on the horse's withers, giving greater leverage. The saddle of the western United States owes its origin principally to this saddle.

Anyone who has seen the Western cowboy cutting cattle or turning a steer has seen how the Moors and Spaniards rode in the Arabian *silla de jineta*. The Spaniards of the islands and the Mexicans on the continent adopted this saddlery. The Westerner rides in the same fashion, in almost the same saddle, and has done so for generations. His saddle is merely the saddle of the *conquistadores* modified by reinforcing and capping the pommel so that it will hold a rope, by redesigning the cantle for added comfort, and by lengthening the stirrups.

The modern Western saddle evolved when the Anglo-Americans took over the range-cattle industry from their Latin predecessors. As they learned their trade from the Mexican ranchers, *rodeos* became roundups, *háquimas* became hackamores, and *da vuelta* became a dally. When having their saddles made, it is hardly surprising that they borrowed any features seeming desirable.

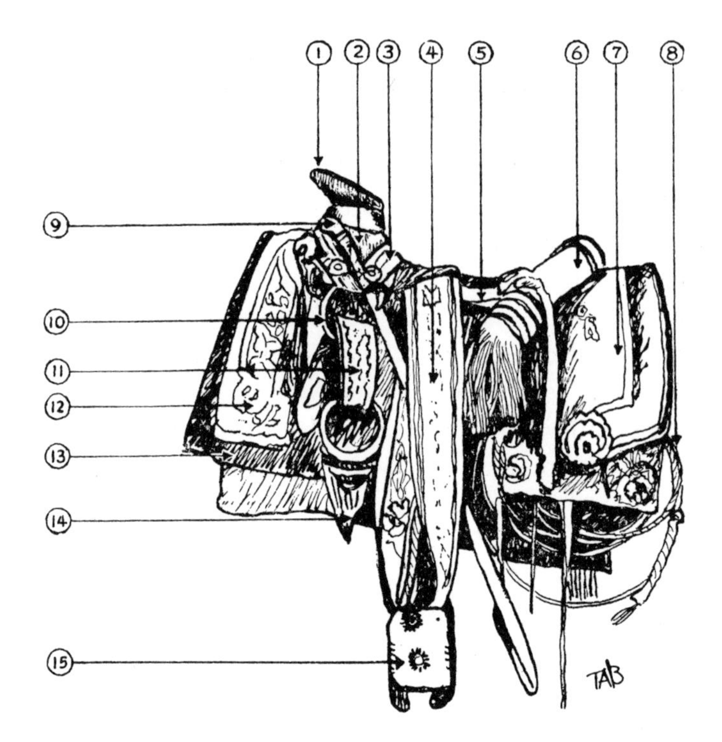

MEXICAN CHARRO SADDLE

Silla mexicana de lujo, bordada de pila con herraje de hierro incrustada de plata

1. Cabeza del fuste
2. Campana del fuste
3. Reatos
4. Arciones
5. Tablas del fuste
6. Sarape
7. Cantinas
8. Reata para Lazar
9. Machete
10. Argollas del enreatador
11. Látigo
12. Bastas
13. Mantilla
14. Cincha
15. Estribo

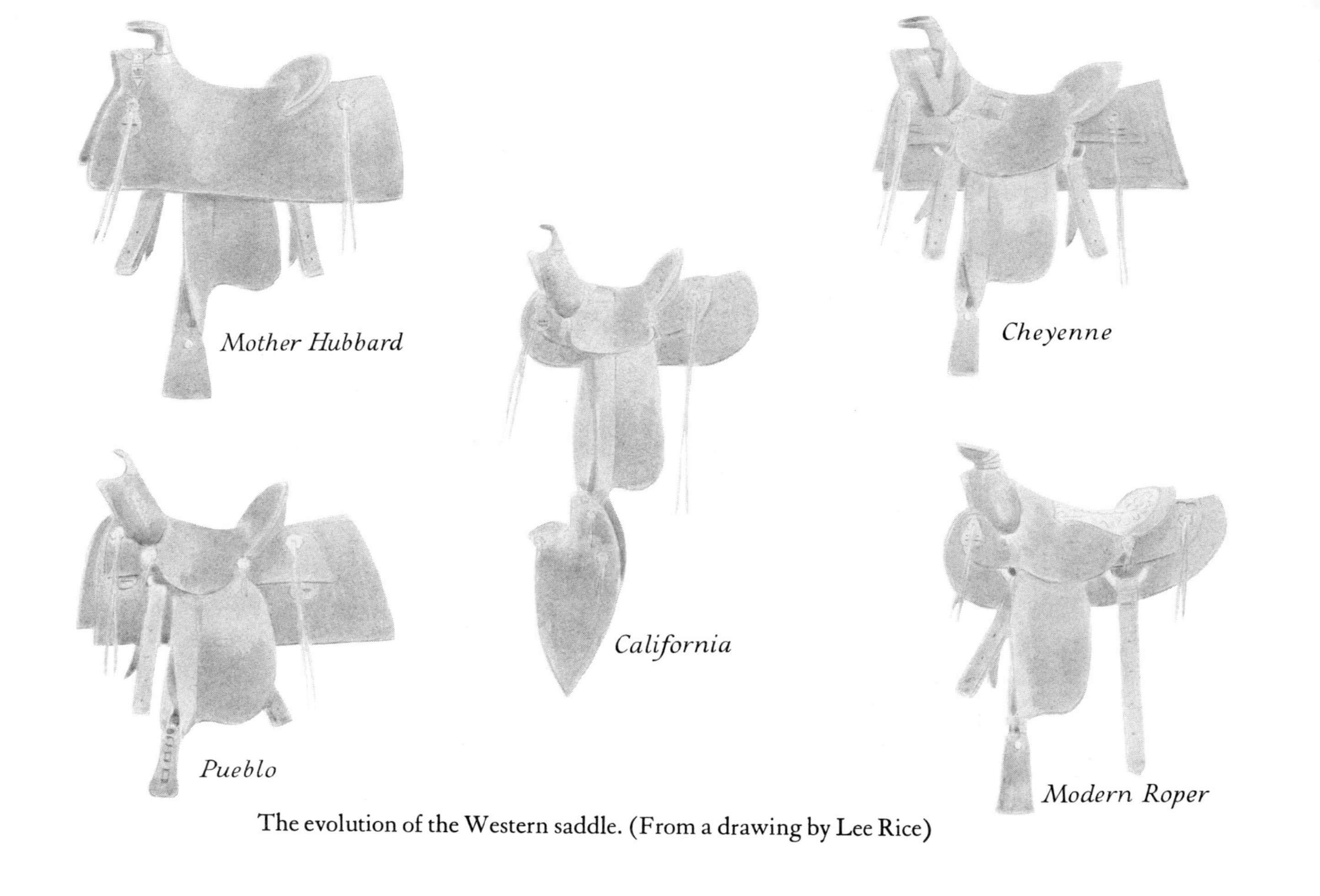

The evolution of the Western saddle. (From a drawing by Lee Rice)

Characteristics and nomenclature of the "Silla Mexicana," or Mexican Saddle

El fuste (tree): This is the foundation of the saddle, made of wood and covered with skin. Its component parts are the *cabeza*, horn; *campana*, forebow; *tablas*, seat; and *tejas*, hindbow.

Los bastos (skirts): The skirts are made of sole leather, lined with undressed lambskin.

Reatas y contrareatas (rigging straps): The *reatas* pass around the forebow under the horn, fastening to an *argolla*, or metal ring, on each side. The *contrareata* runs around the rear bow, also tying onto the *argollas*.

El cincha (cinch): The cinch was generally made of *algodon*, cotton rope, or *cerda*, horsehair, ending in a *hebilla*, or fastener, in the form of a ring which matches the ring on the rigging straps.

El látigo (cinch strap): The strap which fastens the *cincha* to the *reatas* or the cinch to the saddle. The strap on the left is the *látigo*, and the strap on the right, the *contralátigo*.

Las cantinas (saddle bags): Back of the cantle over the *contrareatas* and *bastos* are hung the *cantinas*, which carry the rancher's necessities.

Las arciones (stirrup leathers): The broad, beautifully stamped leathers which support the stirrups. They pass over the *tabla*, hanging on each side.

Estribos (stirrups): Made of metal, covered with hide, often inlaid with precious metal.

Mantilla (saddle blanket): The saddle blanket matches the color of the *sarape*, its edges being visible around the borders of the *bastos*.

Reata de lazar (lariat): The lasso rope is tied below the *catina* on the left side.

Sarape (cape): The *sarape* is the Mexican's raincoat of cloth beautifully woven in many colors.

Machete (swordlike knife): Another indispensible article is the *machete*, which, similar to the American rifle, is hung under the left *arcion*.

Probably the earliest type was the Mother Hubbard, evolved in Texas in the middle of the nineteenth century. It was a low and open saddle somewhat resembling that of the pony-express rider. The covering was generally a single sheet of leather, with the low cantle and right-angle horn projecting through. There was no swell on the forebow and the overall simplicity of the saddle was valuable when saddle shops were few and far between. By the time of the trail drivers it was almost universally used by the livestock men and the cowboys east of the Rockies.

Far to the west in California, still another saddle evolved, soon after the appearance of the Mother Hubbard in Texas. The California creation had two distinctive features—the long tapaderas and the unobtrusive rounded skirt. The forebow and cantle were a compromise between the Mother Hubbard and the Pueblo. This saddle gained immediate popularity in the West and soon was the standard outfit in California, Nevada, and Oregon, and rather popular in Arizona and Idaho. Its modern versions are still common.

A few years later another type appeared and gained favor in the Rocky Mountain region. It became known as the Cheyenne saddle, and in basic outline was similar to the Mother Hubbard. However, the single overall skirt of the Texas saddle had disappeared, and more of the rigging came into view. The double skirts, special seat leather, and the Cheyenne roll on the back of the cantle made this saddle distinctive. Any rider who has had a half-broken horse catch him unaware has had impressed upon him the advantage of the wide-backed cantle.

Late in the eighteen seventies still another popular variation arose on the northern rim of the southwestern cattle area. It was referred to as the Pueblo saddle, and was still common in the early years of the twentieth century. The outstanding characteristics were the small, upright nickel horn, the large square skirts, and the high, dished cantle. No Western stock saddle showed less Spanish influence than the Pueblo.

These four distinctive saddle types developed by the Anglo-American cattlemen provide the basis from which all modern saddles are designed. The only other saddle exerting any marked influence is the Mexican.

During the last half century only two saddles sufficiently original to merit comment have appeared. They are the modern roper and the modern cutting saddle. Designed functionally, the roper is standard equipment for rodeo-contest cowboys, and essential for the doggers and ropers. Because of its comfort and all-around utility, it is found increasingly on the range. For the modern roper an especially low tree was designed to serve two purposes, to withstand heavy and consistant roping and to provide maximum freedom of movement in, out, over, and around the saddle. The skirt of the roper is in the California style, the round skirts making the saddle lighter and easier to throw into a car and "pack" from show to show. The cantle of the saddle borrowed the Cheyenne roll, which makes the saddle faster for mounting and dismounting and makes moving out of the way of a bad throw easier. From the Pueblo saddle came a modified form of double rigging. A quilted seat is customary, not for comfort but for greater security. The thick, stubby pommel reflects the Mexican influence.

Except for being a little more undercut in front and having a smaller cap on a shorter horn, the cutting saddle is similar to the roper. Both customarily have flank cinches. The amazing growth in popularity of the Quarter Horse has also resulted in some changes in saddles. The low, broad withers of the Quarter Horse have resulted in a saddle designed to sit and ride better on the typical Quarter Horse back.

Leather stamping, special designs, silver mounting, and other nonfunctional features were incorporated on any or all of these saddles, depending on the desires and pocketbook of the person buying or ordering a saddle. Almost every cowman has certain features he wishes embodied in his saddle—a good reason why almost no two are identical. Any article in such constant use and so essential to the everyday performance of duty could not but reflect the personality of its user.

☙ *Western Brands*

Although one generally thinks of brands as relating to cattle, they are equally important for horses. As a rule stockmen use the same brand for both animals, although this is not, and has not, always been the case. The same statement could be made about the location of the brand. Of late years there has been an increasing tendency not to disfigure a horse's body with the usual large brands. Some stockmen have begun to brand more on the jaw, on the neck, under the mane, or on the rump, where the brand is partially concealed by the tail, in this way eliminating unsightly marks. The Argentines have perhaps the best method. They, in many cases, brand on the back, so that the saddle blanket covers the brand when the horse is being ridden.

Brands often include more than just the fire brand. Earmarks are almost equal in importance in marking stock, and a stockman can often recognize his stock more rapidly by his earmarks than by his brand, particularly when the hair is long and the brand is old. The most common earmarks are crops, swallow-forks, under-slopes, and under-bits. Those less common but used are over-bits, jingle-bobs, flip-flops, ear-holes, under-sevens, and half-crops. There are other distinguishing marks such as wattles, dewlaps, and buds. Dewlaps are made by cutting down a strip of skin on the brisket, wattles by cutting down a strip on the jaw bone, and buds by cutting down a strip of skin on the nose.

The fire brand is the one which has colored Western history with many romantic tales and given to the West a heraldry of its own. Its importance has not declined. It has become an important asset, a trade-mark guaranteeing honesty in weight, breeding, and all business relations.

A brand may be any single letter, figure, insignia, or combination of any or all three. Normally brands are used only in five positions: either naturally, upside down, at a forty-five-degree angle known as "tumbling," horizontally and termed "lazy," or, finally, reversed. Brands are always read from top to bottom and from left to right. When letters are joined, they are called "connected."

The Mexicans and Spaniards had many strange insignia brands which have supposedly come from the rubrics in their names, and for them, since no name was signed correctly without its owners' own rubric, a brand definitely claimed an animal. Americans have often chosen picture brands such as a key, hashknife, bottle, slipper, hat, rolling pin, gun, rake, domino, frying pan, pipe, fish hook, and so on. Some of the most famous brands have been very simple, such as the *flying W* of the King Ranch, or the *101* brand.

Normally an animal is branded on his left hip. One old explanation for this position is that an animal usually falls on the right side, thus exposing the left hip to the iron. Some outfits have completely ignored the hip and run large letters or figures over a whole side. One ranch had a line running from brisket to tail on the left side.

Branding irons fall roughly into two classes: one called a running iron; the other, a stamping iron, generally termed just an "iron." A running iron resembles a straight poker, and the brand is drawn on the animal with the red-hot iron as an artist uses a brush. The ordinary iron has the brand cast on the end so that it merely needs to be pressed against the hide to leave the correct mark or brand on the animal.

The earliest California brands were those given to the first presidios and missions established in California. The presidios all had similar brands, being *Ja, 2a, 3a, 4a,* and *5a,* the "*a*" in every case being connected. The original manuscript order designating the brands for the soldiers' horses and mules has been preserved. The *J connected a* was given to Loreto, now in Mexico; San Diego, the *2a*; Monterey, the *3*a, San Francisco, the *4a*; and Santa Barbara the *5a*. Most of the early mission irons utilized either letters of their name, like *SD connected* in the San Diego de Alcalá or the *CAP connected* of San Juan Capistrano; or used some figure with religious significance such as Santa Barbara or La Soledad. The early California ranchers had many types, including rubrics, figures, and letters. Examples would be Andronico Sepulveda for the rubric; Mariano Vallejo (connected hearts) for figures, and *TOL* (connected letters) for Thomas O. Larkin.

In modern times all the Western states record brands. Montana records brands in ten places on the animal: the jaw, neck, shoulder, side, and hip, on each side of the animal. In Colorado, it is also possible for ten men to own the same brands, each branding in a different place. In practice, the cowman would buy all ten places so that he could keep his own brand. In Texas, brands are recorded by counties. In Colorado and Montana countless different brands are registered, but if the number were divided by ten, the figure would be close to the actual number of owners. Montana has been recording since 1878, and in 1921 there had been about 85,000 brand certificates issued, with over 60,000 brands apparently in active use.

There are several instances in Colorado of more than one person's owning the same brand. There were ten legal owners of the *O* brand in 1921. In case of disputed animals with no additional dewlaps, earmarks, or wattles, the ranch closest to the shipping point is considered the owner. This duplication of brands occurs because at one time each of the counties had authority to register brands. In Nevada, the brand system provides that the brand be recorded in the county of the owner's residence and then filed in all adjoining counties. In Utah, the records are cleared of dead brands by the requirement that all brands be recorded anew every ten years. In Oregon, by a law passed in 1915, the state veterinarian became recorder of the brands. In South Dakota the brand law was adopted in 1898, and until the recent check there were some 11,895 brands on record. A Livestock Sanitary Board was created in Arizona in 1898, and all brands were transferred at that time from county to state books. The New Mexico State Brand Department has well over 40,000 brands recorded to date.

Branding is as old as cattle raising, and without it the millions of head of livestock annually turned out on the government ranges, forest reserves, and unfenced grazing areas of the West could never be returned to the proper owners. Brands are so important in cattle states that inspectors guard every road and visit all shipping centers. A good brand inspector is difficult to get because reading brands is quite a trick. To the old range man it became second nature, and today it is a sight to see a brand inspector seated on a fence of a big corral at the stockyards checking animal after animal as it passes by. Top men carry between twenty-five and thirty thousand combinations in their heads, can read all the intricate brands correctly, and know to whom each belongs.

Indian Horse Breaking

The American Indian, in his contact with the horse, showed great ability to domesticate and utilize him to his own advantage. For people unacquainted with any similar animal, the Indians mastered the art of horsemanship with surprising speed. Europeans and Asiatics, accustomed to the horse for countless generations, raised horsemanship to the level of an art and a science, but they were not superior to the Indian in any important respect, particularly when the horse is considered solely from the standpoint of utility.

It was the Indians' custom to attribute a high degree of rationality to animal life, and their attitude toward their mounts was no exception. Their relationship to the horse was one which approximated human companionship. This feeling is not unusual for horsemen. The same feeling is found among the Arabs, the Gauchos, the cowboys, and other groups who have had to live day after day with and on their horses. This relationship is reflected in the customary method used by the Indians in breaking horses. They were firm but, for the most part, seldom cruel.

When the Indian had horses to be broken, the animals were rounded up and a chosen horse roped. Several men were needed to hold the plunging horse. The first step was to accustom the horse to the Indian halter, which was very similar to our hackamore. One man worked towards the horse on the rope. This had to be done slowly, so that the horse could see he was not going to be hurt. The Indian approached the horse cautiously, continuously grunting and talking to him. Chief Long Lance writes that horse talk is a low grunt which seems to charm the horse and make him stand motionless for a moment and listen. It sounded like "hoh, hoh," uttered deep in the chest. The cowboy says the same thing today, only he spells it "whoa."

When the horse became accustomed to the man approaching on the rope, the Indian would then take a blanket and wave it. The first shake of the blanket probably gave the Indians holding the rope a free ride. However, as time went on, the horse would become used to the blanket.

The hardest job for the horsebreaker was to get his hands on the horse. By approaching slowly and talking continually, he was finally able to place his hand on the horse's nose. Wild horses have always had a deadly fear of human contact and have to become used to the human scent before they will allow anyone to touch them. Once the horse has had his head touched, the next job is to cautiously put on the halter, or more correctly, the hackamore. The native hackamore was made of a narrow strip of rawhide, no larger than a leather bootlace. It was looped over the horse's nose and run up back of his ears and then down the other side and through the loop. The slightest pull by the horse would tighten the halter. It was strong and slender and put great pressure on the nerves of the neck and nose, and the horse after one or two pulls learned to behave. The Indian, once he had the halter on the horse, no longer needed extra help on the end of the rope.

The warrior would continue handling the horse, hissing at him, caressing him, and generally getting him to realize that he was not going to be hurt. He would run his hands over every inch of his head, and neck, working down to the shoulders and back. When the area around the flanks was reached, the horse would undoubtedly begin to object more strenuously, and the Indian would have to give a few jerks on the line to make him stand still again. When the warrior had succeeded in running his hands over both sides of the horse's body, he would again get his rope and gently strike the horse with it. This was continued until the horse saw that he was not going to be hurt.

The most difficult task was now at hand. The warrior began to work his hands over the horse's legs. A few well-chosen pulls on the rawhide halter were undoubtedly needed to make the horse stand. Not an inch of the horse's body escaped being touched and rubbed. When this job was completed, the horse was nearly broken, as he was no longer afraid of a blanket or a man's touch. This particular feature could well be adopted by many modern horsebreakers. It is the American custom to saddle and ride the horse by sheer force, while he is still afraid of the man, the blanket, and the saddle. Many so-called broken horses are ridden weeks before they allow the familiarity the Indian insisted on at the beginning.

The actual mounting and riding were the least difficult of the breaking job for the Indians. The man would walk to the side of the horse and press down on the horse's back lightly. He kept pressing a little harder and a little harder. Finally, he would put his elbows across the horse's back and draw his body up an inch from the ground, putting all of his weight on the animal. If the horse showed fright and jumped, a jerk on the thong and the repetition of the act made him stand.

Slowly but surely the brave would then pull himself farther and farther up until he could slip his leg over and mount. Surprisingly enough, few horses broken in this manner offered to buck. If one did, it might be called a case of "passing the buck." The horse would usually stand perfectly still and, after a few moments of petting and urging, would trot off at an aimless and awkward gait.

This was the customary Indian method of breaking horses. The Indians did not often have saddles unless they lived near white settlements. They rode, therefore, on a pelt cinched around the horse. Their riding bridle was just a thong looped around the lower jaw of the horse. The horse was guided primarily by the rider's knees. The Plains Indians almost invariably kept a neck rope dragging when they were riding so that if they were unseated, they could grab the dragging rope and retain their horse. A man afoot on the plains was in a serious predicament.

⁂ *Beginnings of the Modern Rodeo*

Although the roundup has been traced back to the tournaments of Medieval Europe, the only logical derivation is found in its almost exact counterpart—the rodeo of Mexico. The jousts held in the Southern states consisted of a certain amount of pageantry combined with contests of horsemanship of a type derived from knighthood. However, they had little relationship to the normal activities of the Western cattle ranch, whose activities have always been the basis of the American rodeo. It was the opportunity of the ranchers to compete in those necessary skills, like roping and riding, which made the rodeo a natural and spontaneous outgrowth of the cattle ranch. It is probably safe to say that a rodeo or its equivalent would have arisen in the Western cow country even without the precedent

Baldy, one of the greatest rodeo horses in history. The rider is Clyde Burke. (Courtesy *The Cattleman*)

of the Mexican ranchers. It was the opportunity for social intercourse and competitive sport. The surprising thing is that it has been accepted so wholeheartedly by the urban population. Today rodeos are held and widely attended in every part of the United States, regardless of whether or not it is stock country. Although all shows are broadly called either roundups or rodeos, each individual show normally has a specific name, such as the Winnipeg, Canada, "Stampede"; the Grangerville, Idaho, "Border Days"; the Kearny, Nebraska, "Frontier Roundup"; the Stamford, Texas, "Cowboy Reunion"; the Cheyenne, Wyoming, "Frontier Days"; the Pendleton, Oregon, "Roundup"; the "Stampede" of Calgary, Canada; the New York "Madison Square Garden World Championship Rodeo"; and the Salinas, California, "Rodeo."

There are many different localities which claim to have had the first rodeo. If the truth were known, probably none of them was first. The first rodeo undoubtedly occurred when a cowboy some place on this earth first claimed he could outride and outrope anybody else—and was called on the spot to "put up or shut up."

The two Anglo-American rodeos which seem to have the best claim for producing the first roundups in the United States are Winfield, Kansas, in 1882, and Prescott, Arizona, in 1888. The 101 Ranch claims to have staged the first rodeo at Winfield in the fall of 1882. The town was planning at that time to have an agricultural fair and in looking for an unusual attraction, got in contact with George W. Miller, owner of the 101 ranch. He had available a group of Texas cowboys who had come with him up the Chisholm Trail, and he suggested that they put on a contest of riding and roping—which proved to be the principal attraction of the show. It was twenty-two years later, however, before the 101 Ranch began its great series of rodeos and displays of Western skills.

The occasion which caused the 101 Ranch to enter seriously the rodeo, or roundup, as the producers preferred to call it, was the National Editorial Association meeting at Guthrie, Oklahoma, in

1905. A huge pasture was fenced off for the events. Shortly after two o'clock the mile-long procession entered the field, led by a cavalry band and old Chief Geronimo who had been brought from Fort Sill for the occasion. Geronimo was followed by a long procession of cowboys, Ponca Indians from the ranch, pioneer wagon trains drawn by oxen, and stagecoaches. After the parade a large group of buffaloes were turned into the arena, and a buffalo hunt was staged. Over one hundred Indians took part in this event. Bronc busting, Indian ball, roping, Indian war dances, and Lucile Mulhall and her trained horse all attracted great attention. As dusk began to settle, a string of wagons was seen coming over a hill south of the amphitheater. The wagon train was promptly attacked by a band of Indians. The burning wagons and yelling Indians caused a feeling of awe to settle on the spectators. From 1905 on, the Miller brothers continued holding their show annually at the ranch, and soon turned it into a traveling exhibition. Even before they did so, Cody's "Buffalo Bill's Wild West Show" had captivated America and Europe.

The most famous of the early rodeos were the "Frontier Days" at Cheyenne, Wyoming, the "Roundup" at Pendleton, Oregon, and the "Stampede" at Calgary, Canada. These are still three very important shows; but the Fort Worth, Texas, Salinas, California, and Madison Square Garden shows are now equally important.

Today rodeo is a big business, with many capable men making a living providing stock and staging a rodeo wherever and whenever there is a demand for one. There is a rodeo somewhere in the United States at least three hundred days out of the year, and during the regular season, which starts in March and ends in October, there are, of course, many held each week end.

A few years ago the participating contestants organized in order to unify the rules and to provide good judges for the various rodeos. Members of the organization were then known as the "Turtles."

Today all cowboys competing at the large rodeos are organized, and there is also an organization of the various rodeo producers.

The usual events encountered in modern competition are: saddle bronc riding, bull-dogging, calf roping, steer decorating, bull riding, steer roping, team roping, wild-cow milking, and bareback bronc riding. In addition to some or all of these events, there are also many other attractions to the modern rodeo. From the opening parade through the regular events, the trick riding and trick roping, the clowns, the trained-horse acts, to the closing act, the modern rodeo presents one spectacle after another.

There have been many great men interested in rodeos, from Teddy Roosevelt to Will Rogers. Will Rogers was more than just interested; he used to participate, particularly in the roping events. Jim Minnick, one of his good friends, said that Rogers could do anything with a rope except throw it straight up in the air and then climb it.

Epic Rides

Feats of strength, endurance, and speed have always pricked men's fancy, and of all the romantic traditions connected with horses, few are more fascinating than the stories of unusual rides. Particularly is this true when those rides have been so essential to the development of our West. Some of the rides are noteworthy for speed, some for endurance, and others because of the dangers involved. A few combined all of these elements, and in them we have the tradition that J. Frank Dobie calls the "Saga of Saddle" reaching its highest point.

During the eighteenth century there were some noteworthy rides made in England. In 1745 the postmaster of Stretton rode a relay of horses to and from London. He made the 215 miles in eleven and one-half hours, averaging about eighteen miles an hour. Four years previously, a certain Mr. Wilde, while at the Curragh meeting in Ireland, had claimed that he could ride 127 miles in nine hours. He easily kept his word, for, using a relay of ten horses, he covered the distance in six hours, averaging about twenty miles an hour. Some years later Mr. Shaftoe, using relays, covered a little over fifty miles in one hour and forty-nine minutes. In 1763 Shaftoe won another bet. He wagered that he could ride an average of one hundred miles a day, using just one horse each day, for twenty-nine consecutive days. Again he won his bet. All told, he used only fourteen horses. On one day he rode 160 miles because his horse had tired the day before. All these English rides, however, were made in a country where there were no hostile Indians and during ideal weather conditions.

The Atlantic coast of America also had its share of riders, Paul Revere being the most famous, though from the point of view of horsemanship his feat was of no import. Lafayette made an interesting ride while helping the Revolutionary forces. In August of 1778, he traveled from Rhode Island to Boston, a distance of nearly seventy miles, in seven hours. He returned in six and one-half hours.

The records made by the riders of western America are by all odds the greatest. Especially in regard to stamina and courage is this true. Here were no roads, and seldom even trails. Only the sun and the stars guided the riders. Over white, towering peaks and deep, burning deserts they pushed their way, with Indians lurking in every thicket and wolves and buzzards just waiting to welcome a mistake. These factors were present in the rides of the Western men, and perhaps it was just these obstacles that made them such skillful riders. They learned, where only one mistake was allowed, and they leave a series of epic rides by a race of supermen.

RAFAEL AMADOR

One of the many amazing Western rides was instigated by General Santa Anna in 1834. It seems that a political scheme had been laid whereby the California mission lands would be seized and sold. A vessel had already left for Monterey with orders to carry out this objective. It was imperative that a counter message reach California first. The general called in Rafael Amador, a young man he knew and trusted, and sent him overland with a counterorder. On the morning of July 25, 1834, Amador started his remarkable ride from Mexico City to Monterey. Some 2,500 miles of mountains, jungles, rivers, and deserts separated him from his destination, but he did not hesitate. Northward across Mexico he galloped, stopping only when he saw an opportunity to change horses. Swimming swollen streams, crossing burning deserts, and climbing tortuous peaks, were all in the day's run. Once in Arizona, he reached the dangerous Apache country. Here he traveled by night. On the Colorado he was ambushed by savages and escaped only with a few clothes and the precious documents. He was now without a horse. One hundred and fifty miles separated him from the nearest settlement, with the present-day Imperial Valley in his path. For days he fought the heat, lying in the shade of the meagre sagebrush when the sun was up, traveling only at night and during the cool of the day. Once over the desert, he struck the hardly more inviting San Jacinto Mountains, but without a falter he trudged wearily on until, parched, tattered, swollen, and torn, he saw San Luis Rey nestled like a mirage in its valley. He had successfully completed the worst stage of his journey. Taking time only to eat and to refresh himself, he was soon drumming north along *El Camino Real*, once again feeling the surge of horseflesh beneath him and listening to the soft creak of leather. Ten days later he arrived in Monterey, giving Governor Figueroa the dispatches on September 11, 1834, just forty-eight days after leaving Santa Anna in Mexico City! Four days later the vessel arrived and the officials aboard found, to their chagrin, a new order countermanding the one they carried. Amador had successfully carried out his mission.

Some years later, John Brown, better known to the Californians as "Slim John" [Juan Flaco] made a grand ride from Los Angeles to San Francisco during the American occupation of California. Brown was with Gillespie, who, with fifty soldiers, had been left in charge of Los Angeles. Gillespie was not particularly successful in controlling either the revolutionary elements of the district or in encouraging friendships, and he soon found himself in hot water. The native Californians, encouraged by the *vino y patriotismo* broadcast during the celebration of *El Grito de Dolores* (the Mexican cry of independence), attacked Gillespie's headquarters. The American was in a difficult position and in self-defense moved from the city to a hill west of town. The natives' healthy fear of the American long rifle kept the patriotic Californians at bay, but Gillespie was still left sitting on top of a hill, six hundred miles from the nearest help.

It was at this point that Juan Flaco volunteered to ride for aid. Gillespie gave Flaco a bunch of tobacco papers, each stamped with his seal and bearing the statement, "Believe the bearer." Flaco left at eight o'clock on the evening of September 24, on the best available horse. He took only his riata and the charmed cigarette papers, both of which were to obtain fresh horses for him before he had gone very far.

No sooner had he left the hill than he was discovered by the watchful Californians. His horse was shot from under him. Carrying his riata and his spurs, so that they would not rattle, he made his way to the ranch of an American living in the Santa Monica Mountains, who, for one of the magic cigarette papers, gave him a horse. On the twenty-fifth of September, at eleven o'clock, he reached Santa Barbara. Again with one of his cigarette papers he procured a fresh horse from Lieutenant Tallbott and was soon speeding on his way. He was discovered by the loyalists and pursued, but outran them by riding his horse to death. Luckily he was near the Robbins ranch, where he got a horse. Later he stopped at the ranch of Louis Burton—this was on the twenty-sixth—and obtained another fresh horse. The following day, only fifty-two hours after leaving Gillespie in Los Angeles, he arrived in Monterey (the modern highway, smooth and straight, stretches 341 miles between the two cities). In Monterey he took his first and only nap of the trip. When he awoke, Maddox had a horse ready, and Juan Flaco pounded on to reach San Francisco at eight o'clock on the evening of the twenty-eighth, having ridden some six hundred miles in four days.

FRÉMONT

Not long after Brown's ride, Frémont, the famous trail-blazer, made the same journey in equally remarkable time. It would be hard to find records showing the strength and endurance of the hardy California horse better than Frémont's ride from Los Angeles to Monterey. Frémont, who was acting as governor of Southern California at this time (1847), feared an uprising of the restless Californians and decided to go to Monterey and talk to Kearny, his superior officer. Leaving the Pueblo de Los Angeles on the morning of March 22, he headed for Santa Barbara. With him rode two companions, and they drove ahead of them in the customary manner six spare horses. Whenever the horses they were riding showed signs of fatigue, they would change. By nightfall they were past Santa Barbara and asleep at the Robbins ranch. When dawn broke, they were on their way again. Sunset that evening saw them eating supper at the Nipomo ranch at Dana, and at nine o'clock they were with friends in San Luis Obispo. The next morning they lingered over a large breakfast and did not leave until almost noon. They were on fresh horses, having left the others in San Luis Obispo. When night fell, they were on the Salinas River, but little sleep was in store for them that night as a group of bears wandered by and frightened the horses. They were again in the saddle by daylight and fanning for Monterey. At three o'clock that afternoon, Frémont walked into Kearny's office in Monterey. Frémont said that he rode 420 miles on this jaunt, although that figure seems a trifle large. At four o'clock the next afternoon, his business completed, he was ready to return to Los Angeles.

Frémont had been given two beautiful California horses at San Luis Obispo for the ride north. Jesús Pico had made the gift. *Los canelos*, as they were known because of their cinnamon-colored coats, were more than equal to the trip. On the return Frémont mounted the older animal and left Monterey, headed south at a good gallop. Colton, the American alcalde of Monterey, who saw him leave, said, "A more graceful horse and one more deftly mounted I have never seen." The "Pathfinder" rode the horse fifty miles that afternoon. The following day he rode him ninety miles without changing. While still a little distance from San Luis Obispo, he shifted to the younger horse, although the elder *canelo* still showed no signs of weakening. When they started again, the horse took the lead and kept it all the way to his home pastures.

A rainstorm delayed the party until nearly noon of the following day, at which time they left San Luis Obispo and once again headed south, with their original nine animals. For the next two days they averaged 130 miles a day, and they reached Los Angeles just nine days after the start. Although only some seventy-six hours had been spent in the saddle, they had covered around 840 miles according to Frémont's estimate. This was a nice jaunt.

FELIX X. AUBREY

If any one man stands out as the supreme rider the West has known, certainly it is the amazing Felix X. Aubrey. It was his unbelievable rides (coming at a time before the Pony Express was dreamed of) that started men's tongues wagging with the idea of quick communication with the East. New Mexico, a land of horsemen, and Santa Fé, a thriving horse mart, were the inspiration for his rides.

In romantic nineteenth-century Santa Fé three races mingled: Spanish, Mexican, and Anglo-American. Love of horseflesh was one common factor. Aubrey's blue-black hair and slight frame elicited no comment, but the horse he rode, a magnificent animal, caused a murmur to run through the loungers on the corner. There was nothing about the appearance of this man, who was only five feet tall and weighed a scant one hundred pounds, which gave any indication that he was soon to electrify the continent with the speed of his trips from Santa Fé to Independence, or to hint that some day he would be considered the greatest rider of the West.

By 1848, the year of Aubrey's three greatest rides, the commerce of the prairies was approaching its greatest tide. A major trunk of this trade was from Independence to Santa Fé, a distance of some eight hundred miles. The trip generally took from two to three months by caravan, although horsemen sometimes made the trip in approximately a month.

One bleak January day Aubrey left Santa Fé and just twelve days later arrived in Independence. Both towns were agog with excitement when the news spread. Little else was spoken of for weeks. Spring had hardly broken when a new tumult arose. Aubrey had said that he was going to make the same ride in eight days. This, replied the townsmen, was impossible: on a well-marked trail with no hostile Indians, perhaps; but across the rough and dangerous uncharted territory separating Santa Fé from Independence, never. Frontier towns in general, and horsemen in particular, are always willing to wager, and the bets ran high—mostly against Aubrey.

Aubrey left Santa Fé with six companions. They were all far behind before the first third of the journey was completed. He rode three horses to death, was attacked by Indians, and lost even the few things which he carried; he walked and ran forty miles; he slept on the ground only 240 minutes; he went three days without food; but he arrived at Independence on the eighth day to collect his bets.

In the fall he was again in Santa Fé. While talking with Kit Carson, he suddenly found himself with another bet on his hands. This time he wagered that he could make the trip in six days, the whole eight hundred miles, provided he could use relays. So sure was he that he put up one thousand dollars on the outcome. The money was instantly covered, and once again the town was talking nothing else but Aubrey.

On the morning of September 12, Aubrey left Santa Fé, and from that moment he apparently did not stop to eat, sleep, or drink, until late evening on September 17, when he rode into Independence, a ghost of a man who could speak only in a hoarse whisper when lifted from his blood-caked saddle.

Through a land that was claimed and controlled by the Kiowas, Apaches, and Comanches, he had swung his way, with only the monotonous song of pounding hoofs and creaking leather to relieve the roar of the blood in his ears. The first day was probably not unpleasant, but after the second day the desire to sleep must have been overpowering. He must have strapped himself to the saddle. Much of the trail was swampy and cut with raging streams. Swimming icy streams at night was no joke. Out of New Mexico, into Colorado, across the No-Man's-Land of Indian Territory into Kansas and Missouri—a solitary man, eight hundred miles, and not a single hour of rest.

JOHN PHILLIP

Somewhere near Cheyenne lies an unhonored and unmarked grave, and in it a courageous and unsung hero—John Phillip. Phillip did not make his ride for money. He was not riding with the government mail for a salary. He was not riding for a wager. A more powerful force drove him on to superhuman efforts, the lives of his comrades and the people he loved.

During the cold, blustery December that covered the West in 1866, the Indians silently surrounded Fort Phil Kearny. For months Red Cloud had been struggling to keep his vow and destroy the fort. Since August he had killed 154 persons, wounded 20 more, and confiscated most of the livestock. Fifty-one times the Indians had openly attacked the fort, and fifty-one times they had been driven back by the plucky but ever smaller garrison. Red Cloud, despairing of capturing the fort by open attack, had led his braves back once more, and now they lay in a grim circle of death surrounding the fort in a relentless siege.

Friday, December 21, dawned; the brilliant morning sun made the snow on the hills glisten and the ice-hung limbs snap and pop in the morning air. Before that sun set in the evening, eighty-one of the remaining soldiers had become pincushions for Indian arrows and lay stark and naked, a strip gone from the top of each head. They had made the mistake of leaving the fort. So weak was the fort now that without help it was merely a matter of time before it fell. Even the ammunition was too low to withstand another attack.

Two hundred miles south and west lay Fort Laramie with warmth, safety, and reinforcements. Two hundred miles, but then it might as well have been two thousand. In some way word must get through the silent red and white death which surrounded the fort. Everyone knew that someone must make the attempt, but who? Then John Phillip stepped into history. He made only two requests. He wanted to speak to a lady and to have his choice of horses. Both were instantly granted.

Choosing a short, heavily muscled animal and swathing himself in a huge buffalo robe, he slipped away into the night. By day he lay concealed and by night he galloped. What he and his horse suffered from the cold and exhaustion, or how he guided himself through the two hundred miles of monotonous repetition that was the snow, no one ever knew. His beard froze solid first, then his hands, feet, and knees froze, but some spark within him kept him moving.

At Fort Laramie the Christmas ball was in full swing. Suddenly above the sound of revelry came the warning shot of a sentry. Music and dancing ceased abruptly. The light from the open door framed the figure of a dying man and horse. John Phillip's body delivered his message.

PONY EXPRESS

The scheme of using relays of horses to facilitate rapid transit of news is almost as old as the domesticated horse. It was under Genghis Khan that the greatest horse express was developed. The Khan latticed his kingdom with a group of horse relays the like of which has never been seen since. However, in the western United States, the Pony Express became famous because of its riders.

"Pony Bob" Haslam was only one of the regular Pony Express riders, and his feats, although remarkable, were no greater than most of the riders could, would, or did accomplish under similar circumstances. Haslam carried the mail from Friday's Station to Bucklands. While the Pah-Ute war was raging in Nevada, his difficulties increased. The distance between the stations was seventy-five miles. One morning Bob arrived at the station immediately west of Bucklands and found that the Indians had attacked the unit and driven off the stock. There was nothing to do but to ride on with the same tired horse. When he arrived at Bucklands, he found no relief there either, and had to continue with the mail. Galloping through Carson Sink, sixty-five miles away, he came to Sand Springs, where he changed horses, ninety miles from his starting point, and continued on his way. Soon he met the west-bound rider and they exchanged *mochillas*, or mail pouches. Since Haslam still had no relief rider, he started back over the trail he had just traveled.

The Indians had struck the Post Station in the few hours since he had gone through. The buildings had been burned and the keeper killed. On reaching Sand Springs, he changed horses and told the keeper he had better come with him, as the Indians were on a rampage. The keeper refused. The next day the Indians struck Sand Springs and killed the keeper. Haslam galloped on and arrived at Bucklands, the mail only a little more than three hours behind schedule. When Haslam had completed the run back to Friday's Station, he had covered 380 miles and had been in the saddle for thirty-six continuous hours.

Eastern Blood

Although the principal basis for the Western horse is found in the Spanish animals imported into Colonial America, in the western United States the Spanish horse has received numerous crosses of the blood brought into the Anglo-American colonies by the English, French, Dutch, and Swedes. Today, almost without exception, Western horses contain some Arabian, Thoroughbred, Morgan, Standardbred, American saddle horse, or some other blood of the heavier breeds such as the Percheron, Belgian, Clydesdale, and Suffolk.

The Arabian horse is generally recognized as the first breed of livestock developed by man. He was taken from Arabia across Africa to Europe by way of Spain and was later imported into America in the sixteenth century and into England at the close of the seventeenth century. In England he helped produce the English Thoroughbred, the Norfolk trotter, and the modern Hackney. The Arabian horses brought to England from the eastern shores of the Mediterranean by returning soldiers together with those imported from France and Spain are considered to be the base of the English Thoroughbred. Generally the Darley Arabian, the Godolphin Barb, and the Byerley Turk, all bearing Arabian blood, are considered the foundation sires of the Thoroughbred. Since most American Thoroughbreds sprang from imported English Thoroughbreds, there is Arabian blood in the foundation of practically all American light horses.

Arabians are all of a solid color, with white and black appearing only occasionally. Their skin is dark and shows black around their eyes, nose, and muzzle. They are normally rather small, standing from 14-1 to 15-1 hands and weigh from 850 pounds to 1,000 pounds. The Arabian possessed for centuries superior refinement, intelligence, docility, spirit, quality, beauty, style, speed, and endurance. The head is one of his most striking features, being clean cut, tapering from the eye to the muzzle, with a depression or dish in the face. The Arabian horse has been raced with some success in India and elsewhere.

One of the first importations of Arabian horses into America by the English people was that of Ranger in 1765. Probably the most significant importations were those made by Homer Davenport and Randolph Huntington.

The Arabian has been bred with care toward a definite type for a longer period than any other breed of livestock, and all the breeds of lighter horses and most of the heavy breeds carry in a greater or less degree the blood of the Arab.

The English Thoroughbred was developed by crossing native running mares with outstanding Arabian stallions that were brought into England. The United States was the first country after England to breed the Thoroughbred. The first American importation was a horse called Bull Rock, in 1730. In the early days Virginia was the leading supporter of the American Thoroughbred and still maintains an important position, although in modern times Kentucky has gained greater renown. Diomed is considered by many to be the most important stallion to be imported into America, and Lexington, one of his descendants, the greatest stallion bred in America. The position of the Colonial Quarter Horse in the development of the American Thoroughbred has already been stated.

The modern animal stands from 15-2 to 16-1 hands and weighs from 1,000 to 1,100 pounds, and his form is long, rather deep-chested, and upstanding. He is often a bit angular, and the degree of style he possesses is moderate. The breed was of course designed primarily for racing under saddle, for which purpose the Thoroughbred has no equal, and the greatest breeders still keep this foremost in their minds.

The Morgan originated in Vermont, and the breed is unique in that it was founded by one horse, Justin Morgan. This stallion was foaled in 1793 and was named after his school-teaching owner. Justin Morgan seems to have been of Colonial Thoroughbred and Quarter Horse breeding. At first the Morgan was used for a light-harness animal as well as for saddle work, but today he is being used primarily for the saddle. Morgan blood was important in the foundation of both the Standardbred trotter and the American saddle horse.

The American trotter or Standardbred originated during the middle of the nineteenth century in the vicinity of New York and Philadelphia in response to the demand for a horse for road driving and harness racing. The breed was evolved from the Thoroughbred, the Norfolk trotter, the Arabian, and certain pacers of mixed breeding. New York, Kentucky, and California were the first breeding centers. Messenger and Justin Morgan were two of the best-known stallions to influence the Standardbred, and Hambletonian was the greatest sire of trotters ever developed. In fact, he was such a great individual that the individuals of the breed are often referred to as "Hambletonians."

In size, the horse varies from 14-2 to 16-1 hands and ranges in weight from 800 to 1,250 pounds. His form is rather upstanding, leggy, long, deep, and narrow. Style and beauty have been secondary to speed. Both trotters and pacers are found in the trotter breed.

While New York and Philadelphia were developing a trotting horse to cover their roads in quick time, Kentucky, Tennessee, Missouri, and Virginia were developing the American saddle horse to use on their foot and bridle paths. The popular American saddle horse is today recognized as a true breed and was created by the fusion of three elements: namely, the Thoroughbred, the Morgan, and the American trotter. All of the early sires and dams were chosen for their easy ambling gait.

Today the saddle horse is traced to two principal families established by a Thoroughbred, Denmark, and to the American trotter, Mambrino Chief. These beautiful animals are divided into two groups, a three-gaited and a five-gaited classification.

American saddle horses are outstanding in loftiness of carriage, airiness of movement, refinement, intelligence, and docility coupled with high courage and spirit. The saddle horse is considered by many as the most stylish, beautiful, and finished of all horses. Certainly his only competitor in this respect is the Arabian. The American saddle horse is notable for his long, refined neck, nice sloping shoulders, moderately high withers, and level flat croup; the tail is set so that it will be carried high. The three-gaited animal has his tail trimmed, while the five-gaited animal sports the full flowing tail. The gaits desired in a five-gaited animal are the walk, trot, canter, rack, and one of the slow gaits, either the running walk or fox trot. The three-gaited horses walk, trot, and canter only. This difference in gait is the only basic difference between the animals.

Color Terminology

When a person grows up around horses, he naturally learns to call horse colors correctly. The color terms used by horsemen are confusing to others, but they are absolutely necessary for correctly identifying a special horse in a group. It hardly seems necessary to point out to the informed reader the variation that exists in the use of color terms across the United States. In order to help clarify and explain the use of color terms, three articles by me that appeared in *Western Horseman* (January, 1948; December, 1971; and January, 1972) have been combined and somewhat rewritten for presentation here.

One is always asking for trouble when he attempts to tell another person what color a horse is. Color terminology is often a personal thing, influenced by the fact that the same color does not always look the same to two people. However, there are many more similarities than differences.

BASE-COAT COLORS

1. White
2. Dun
3. Sorrel
4. Bay
5. Brown
6. Black
7. Grullo
8. Grey
9. Roan
10. Pinto

CHARACTER OF BASE COAT

1. Ratty
2. Parched
3. Toasted
4. Smoky
5. Dark
6. Zebra
7. Line-backed
8. Striped
9. Dappled
10. Flea-bitten
11. Red-speckled
12. Patched
13. Spotted
14. Calico
15. Light
16. Golden
17. Pure

PECULIARITIES OF HEAD AND POINTS

Head:

1. Spot
2. Star
3. Stripe
4. Blaze
5. Bald face
6. Snip
7. Race
8. China-eyed
9. Glass-eyed
10. Cotton-eyed
11. Blue-eyed
12. Watch-eyed
13. Wall-eyed
14. Orry-eyed
15. Mealy-mouthed

Feet and Legs:

1. Socks	5. White points	9. Feathered
2. Boots	6. Striped	10. Smooth
3. Stockings	7. Spotted	11. Clean
4. Black points	8. Coarse	

Mane and Tail:

1. Flax	3. White	5. Broom-tailed
2. Silver	4. Rat-tailed	6. Bang-tailed

Your friend has bought a new horse. Your first question is, "What color is he?" The answer might be, "A dark sorrel with a blaze and one stocking." His answer includes all three of the common descriptive categories for color markings. "Sorrel" denotes his basic coat color; "dark" describes the character of the base-coat color; "blaze" and "stocking" indicate peculiarities of the head and legs. By using just adjectives from each of these categories, one can describe any horse so that he can be easily identified.

In the above list category A represents base-coat colors. The second category, B, represents shades and variations of the general coat color, and the terms are used as modifying adjectives for category A. Category C denotes peculiarities of color in the head, legs, mane, and tail, and these terms are the key adjectives for identification when large groups of horses are under consideration.

BASE-COAT COLORS

The first, and most important category is, of course, the basic coat-color terms applicable to all horses. When only a few horses are being spoken of, this one term is all that is commonly needed or used. For example: "That bay," or "That sorrel."

For the purposes of this discussion there are ten basic color coats: (1) white, (2) dun, (3) sorrel, (4) bay, (5) brown, (6) black, (7) Grulla, (8) grey, (9) roan, and (10) pinto. Almost all of these have general and special modifying adjectives describing the character of the coat color.

1. *White*. The true white horse, an albino, is born pure white and dies the same color. Very little if any seasonal change takes place in his coat color. Age does not affect it. Many have white eyes pink skins, and white hoofs. If, during the life of a white horse, hairs of color other than white are found, the chances are that the horse is not white, but grey or roan.

2. *Dun*. The dun horse is one whose dominant hair is some shade of yellow. A dun horse may vary from a pale yellow to a dark canvas color with mane, tail, skin, and hoofs grading from white to black.

The lightest dun horse is generally called a cream, or, in Spanish *huevo de pato* ("duck's egg"). This horse has just missed being an albino, and his body seems a uniform pale yellow. Often eyes, skin, and hoofs, as well as mane and tail, are yellow, hazel, or amber-colored. Next in line comes the palomino, the best examples of which are a beautiful gold which shines in the sun. Palominos always have a white mane and tail. Then come those which are dappled and darker. Some have black hair in their manes and tails, which are then called silver. The duns range on through yellow to orange to a rich oak color and even darker until they reach a dirty canvas color.

Many duns have black points, and with the development of the Palomino Horse association it seems that common usage is changing "dun" to mean a horse with black points and "palomino" to mean a dun with white points. Special names are also given to these darker duns. A coyote dun is one with black points and a line down the back. A zebra dun is one with black points and zebra stripes on his legs and shoulders. A red dun is a dun of reddish-orange cast, often with a red stripe down his back and a red mane and tail. In the Thoroughbred studbook these horses are listed as sorrels, and sometimes ranchers call them claybanks.

To summarize: Dun is a horse color in which yellow hair predominates. If it is light and mixed with white, the dun is called a *huevo de pato*, cream, or cremolo. These horses never have black points. If the coat is pure golden, the horse is a palomino. They always have white points. If the coat is yellow, mixed with red and has red points, he is a sorrel or claybank. If the coat is yellow mixed with bay and has black points he is a buckskin. A dun horse whose coat is mixed with brown or black is a grullo (discussed under a separate heading).

3. *Sorrel.* A sorrel is a horse whose coat is basically red. His mane and tail are normally the same shade as his body. If the mane and tail have some white in them, the horse is termed a flax sorrel. The mane and tail of a sorrel horse is never black. A light sorrel is one that is a bright yellowish-red. A dark sorrel, commonly designated in the west a chestnut sorrel, is a deep, dark, rich mahogany red. All sorrel Thoroughbreds are called chestnuts. Some are so dark that when they are standing in the shade it is difficult to tell what color they are. They may look almost black. However, by looking carefully at the mane and tail or at the fine hairs around the nose, the red may be detected. A liver sorrel or chestnut is dark; a red or golden sorrel, light.

4. *Bay.* A bay horse is one whose color is hardest to describe but easiest to distinguish. It is a mixture of red and yellow, probably best described as much the color of a loaf of well-baked bread. A light bay shows more yellow, a dark bay more red. The darkest is the mahogany bay, which is almost the color of blood, but without the red overtone. Bays always have black points, occasionally white feet. A red bay can never be confused with a sorrel as bays always have black manes and tails; sorrels always a red (or occasionally flax) mane and tail. The body color of a mahogany bay and a chestnut sorrel can be the same, but the mane and tail provide an easy method of identification.

5. *Brown.* A brown horse is one whose coloration is, curiously enough, brown. Many brown horses are erroneously called black because they are so dark. A close examination of the hair on the

muzzle and around the lips will quickly tell whether the horse is brown or black. The mane and tail are always dark.

6. *Black*. A black horse is black, and almost unvariably has black eyes, hoofs, and skin. The points are always black.

7. *Grullo*. The grullo always has black points. He is a smooth, greyish-blue color like a mouse. He is not a blue roan, however, or a grey. Some horses seem almost purple or smoke-colored. Most of them are line-backed and have zebra stripes on the legs and withers. They are not classified here as duns because the dominant coloration appears blue or mouse instead of yellow, as is the case with the true dun.

8. *Grey*. Most so-called white horses are really grey. Many people even call an old grey horse an albino, especially if it has light skin and hoofs, and one or more white eyes. The normal grey horse is born blue or almost black, and, as he grows older, more and more white hairs come into his coat, and the dark hairs are shed, until by the age of eight or ten most appear white.

The beautiful dapple-grey color generally comes between the second and fifth year. Young grey horses look like and are often spoken of as blue roans. When small specks of black hairs are present, the grey is called flea-bitten or blue-speckled. When a grey horse is young and has a great deal of black in his coat, he is termed iron-grey or, if the black makes a pattern, dappled-grey; a little later, when more white shows, he may be called a silver-grey.

9. *Roan*. A roan horse is any horse whose coat carries white hairs intermingled with one or more base colors. Many are born and die about the same color. Any roan can be light or dark, depending on the proportion of white hairs compared to the colored ones. Most roans are combinations of bay, sorrel, or black, with white hairs intermingled. They are known, respectively, as red, strawberry, or blue roans. The roan coloration is generally not uniform, and some patches on the body will be darker than others. A roan is not a pinto, although most overo pintos are roans. Some roans

are predominantly white. They are known as flea-bitten or blue-speckled if the small spots are black, red-speckled if the spots are bay, and *sabinos* if the spots are sorrel or red. A grey horse is a roan, but not all roans are grey.

10. *Pinto.* A pinto is any horse that has more than one color in or on his coat in large irregular patches or spots. Small nonwhite spots, up to the size of four bits, embossed on a color other than white do not necessarily indicate a pinto. For example, many sorrel horses have small black spots on their rumps. A great deal of white on the upper legs or face is a pretty good indication of pinto blood, as is any white spot above the knees and hocks or outside the rectangular area on the face outlined by the ears, eyes, and nostrils.

Pintos can be classed in at least three categories. Two of these categories (from the South American) are the *tobiano* and the *overo*. The third are the spotted horses, such as the leopard and the Appaloosa. A *tobiano* is a cleancut pinto, often called a spotted horse, showing no roan hairs. Normally having white as the base color *tobianos* may have large blocks of dun, sorrel, bay, brown, or black on the body. Most *tobianos* have a normal head color. Their mane or tail is the same color as the coat at its source. If two colors meet on the neck, two colors are found in the mane, but each distinct and clearly bounded, depending on its origin. Some *tobianos* have more than one base-coat color superimposed on white.

The *overo* is what many Westerners call a paint, a patched, a calico, or an Indian pony. The *overo* is a roan pinto. The color blocks are normally smaller and not clearly defined as in the *tobiano*. As an aid in identification, consider that a *tobiano* is never a roan, while an *overo* is always a roan. Unfortunately, occasionally the horses combine in some form characteristics of both.

A leopard-spotted horse is one whose base coat carries a multitude of small spots. An Appaloosa is the same, except that the spots generally seem to be centralized around the rump and hip. The Appaloosa may be otherwise a solid-colored horse or an overo pinto or roan.

A piebald is a *tobiano* pinto, which is white and black. A skewbald is a *tobiano* pinto which is bay and white or sorrel and white.

The *overo*, or roan paint, as the horse used to be called, may be almost any color. He may be grey, dun, grulla, sorrel, bay, brown, or a blue in combination with white.

CHARACTER OF BASE COAT

When discussing a group of horses, and there are several of one color present, it often becomes necessary to distinguish between two of the same base color. One way is to use an adjective which gives specific character to the common base color. There are any number of these modifying adjectives, but seventeen are enough to cover most any situation:

1. *Ratty* indicates lack of uniformity in color—a dull, dirty tone.
2. *Parched* means faded, washed out, yellowish, or dried up.
3. *Toasted* implies darker patches, dull finish, or a dark overcast.
4. *Smoky* means a blue tinge to the color, an obscure tone.
5. *Dark* indicates a predominance of black hair or deep color, with little yellow apparent.
6. *Zebra* always means dark stripes on the legs and/or withers.
7. *Line-backed* indicates a darker ribbon that goes along the dorsal region from mane to tail. The line may be almost any color, although red and black are most common.
8. *Striped* indicates black stripes or bars on the legs.
9. *Dappled* means darker spots embossed on the coat.
10. *Flea-bitten* is a grey or roan horse having small black or blue specks or spots on a predominantly white background.
11. *Red-speckled* indicates a grey or roan horse having small bay or sorrel specks or spots on a predominantly white background.
12. *Patched* indicates large roan spots on some base color.
13. *Spotted* indicates spots of solid color on some base coat.
14. *Calico* is the same as patched, although generally applied to more lively color combinations, normally applied to an overo pinto.
15. *Light* indicates a predominance of yellow or white hairs.

16. *Golden* refers to the sheen which, when the light strikes certain shades of dun, sorrel, and bay, makes them seem translucent and golden.

17. *Pure* indicates uniformity, clarity, and depth of color; it is generally a sign of a healthy horse.

PECULIARITIES OF HEAD AND POINTS

When discussing or describing an individual horse among many, it is necessary to be more explicit than merely to use a general color term with a modifying adjective. Instead of saying "Dark sorrel," it may be necessary to say, "Dark sorrel with a blaze face." There are two general ways of doing this. The first is by using special descriptive adjectives applied to the head. Among the many used the following are common:

Head:

1. A *spot* is a small patch of white on the forehead about the size of a nickel.

2. A *star* is a patch of white on the forehead about the size of a quarter or larger.

3. A *stripe* is a long, narrow band of white working from the forehead down toward the muzzle.

4. A *blaze* is a wide stripe.

5. A *bald face* is one in which white covers most of the flat front surface of the face, often extending toward the cheeks.

6. A *snip* is a small patch of white which runs over the muzzle, often to the lip.

7. A *race* is a small patch of white running across the face. It is always at an angle.

8 to 14. Normally horses have a rich brown eye with a black pupil, and no white shows around the edge. When this coloration varies, many adjectives are used to distinguish the difference. When

the eyeball is clear, some shade between white and blue, he is normally termed *china-eyed*, *glass-eyed*, *cotton-eyed*, or *blue-eyed*. If only one eye is light, he is often termed *watch-eyed*. If one eye is defective, he is called *wall-eyed*. (In some places *wall-eyed* refers to the white of the eye showing or to the white in the face covering the eye area). The term *orry-eyed* is also used to denote a horse who shows, because of fright or because his pupil is overly contracted, white around the rim.

15. A *mealy-mouthed* horse is one whose color fades out around the mouth, found in bays and browns especially. Occasionally this characteristic is called mulish because so many mules are mealy-mouthed.

Feet and Legs:

The feet and legs are equally important for identification and description as head characteristics:

1. A *sock* is white, extending from the hoof to the pastern or fetlock.

2. A *boot* is white, extending from the hoof into the lower half of the cannon bone.

3. A *stocking* is white, extending from the hoof above the middle of the cannon bone toward or to the knees or hocks.

4. *Black points* indicate a mane and tail (and generally the lower legs) that are black.

5. *White points* indicate a mane and tail (and generally one or more feet) that are white.

6. A *striped* horse is one with black stripes on his legs and/or withers.

7. A *spotted* horse is a pinto or roan with spots of different color on the body or upper legs. White on the face or lower legs does not indicate a spotted horse.

8. *Coarse* legs are the opposite of clean legs. Although normally

caused by the physical formation of bone and tissue, the coarseness is accentuated when the hair lacks uniformity and is shaggy and rough.

9. *Feathered* legs are those on which long hair grows on the backs of the lower cannons or fetlocks.

10. *Smooth* legs, although basically caused by physical formation, are accentuated when the hair is short and uniform.

11. *Clean* legs, like smooth and coarse legs, are not primary adjectives used when referring to the coat; however, clean legs are almost never shaggy, rough, or coarse in regard to hair growth.

Mane and tail:

Black points always indicate a dark mane or tail, while white points or light points mean a light mane or tail.

1. *Flax*, when applied to mane and/or tail, indicates a straw yellow or dirty white. It is normally caused by a mixture of dark color with the white.

2. *Silver* denotes a mane or tail which is white with a little black, giving it a silver cast.

3. *White* manes and tails have only white hairs.

4. A *rat-tailed* horse is one having but little hair in his tail. Many Appaloosas have this characteristic.

5 and 6. A *broom-tailed* or *bang-tailed* horse has a heavy, coarse tail. Many mustangs have this characteristic.

If the three above categories of color terms are learned and used properly, no one needs to worry about his ability to describe or identify a horse properly.

☙ *Latin-American Color Terminology*

My experience with Spanish color terms is limited to those used in South Texas, Mexico, and Argentina. Other countries (as well as sections within these three countries) have somewhat different classifications. However, there are enough common factors to make the following lists useful for anyone traveling in a Latin-American country or for translating from the Spanish.

TERMS OF THE HEAD, MANE, TAIL, AND LEGS

1. *Estrella.* A star on the forehead (a small white mark).

2. *Pico.* A snip on the nose (a small white mark near the mouth).

3. *Lista.* A stripe on the face (a long white mark).

4. *Malacara.* Bald-faced, or having excessive white on the face (sometimes called *pampa* in Argentina).

5. *Antiojeros.* Black rings around the eyes.

6. *Ojos negros.* Darkly pigmented eyes.

7. *Zarcos.* Glass eyes or watch eyes (the eyes lack dark pigmentation).

8. *Cabos negros.* Black mane and tail.

9. *Calzadas.* Sox (white markings on legs to pasterns).

10. *Calzadas medianas.* Stockings (white markings on legs to knees or hocks).

It soon becomes obvious to the North American that the Latin American's use of color terms is quite different from his own. He finds that a color that he has always considered a basic coat color is simply a modifying adjective for the South American horseman. The North American *sorrel*, for example, can be a *bayo*, an *alazán*, or a *castaño*. The South Americans use *alazán* to refer only to horses whose coat color is yellow. All of what we call chestnut sorrels or liver sorrels are *castaños*. On the other hand, some of our lined-back, or very pale-yellow, sorrels may be a type of *bayo* in Latin America. To avoid too much confusion for the beginner, some of the more commonly used color terms and a few modifying adjectives to go with them are listed below. Using them as a base to start from, the reader can gradually learn the Latin-American terminology. Incidentally, to me, at least, Latin-American usage seems more accurate than ours. This is especially true with terms like *bayo*, roughly translated as "dun." To the Latin American *bayo* covers a whole group of colors, which change as the dilution factor increases. They start with a dark, zebra-striped dun and gradually fade out to the palomino, the cremolo, and the albino.

The following lists have been generally organized by listing the main terms first and the more common modifying adjectives in the second list. No attempt was made to list basic color terms separately.

COMMON TERMS

1. *Alazán*, sorrel. Includes only those sorrels with yellow hairs, not chestnuts or red sorrels:
   Claro, light
   Tostado, dark
   Dorado, gold
   Requemado, brown or sunburned
   Tipo, all hairs, mane, and tail the same color

2. *Bayo*, dun:
   Blanco, light
   Amarillo, yellow
   Naranjado, orange
   Encerado, wax
   Dorado, gold
   Atigrado, tiger-striped
3. *Blanco*, white:
   Plateado, silver
   Mosqueado, flea-bitten
   Albino, pink-skinned
   Rosado, white with a few bay hairs
   Porcelano, white with blue spots
   Paloma, yellowish
   Sucio, dirty
   Azulado, blue

4. *Cebruno*, dark zebra dun. Always black skin, hoofs, and points. Sometimes confused with *gateado pardo.*

5. *Colorado*, bay:
Claro, light
Pardo, dark
Requemado, burned

6. *Gateado*, dun, with or without zebra markings:
Claro, light
Pardo, dark
Rubio, red or bay hairs
Barcino, ruddy brown

7. *Huevo de pato*, faded light yellow, a washed-out color. Represents progressive albinoism.

8. *Lobuno*, blue dun, grullo, no zebra stripes

9. *Moro*, blue roan:
Claro, light
Escuro, steel-colored

10. *Tordillo*, grey:
Blanco, light
Negro, blue
Plateado, silver
Mosqueado, small black spots
Sabino, small red spots
Rucio, frosted

11. *Zaíno*, brown:
Claro, light
Negro, dark
Pardo, dark
Colorado, red

GENERAL MODIFYING ADJECTIVES (can be used with most colors in list above)

1. *Azafranado.* Any horse born one color that changes color with age.

2. *Cebrado.* A horse zebra-marked on the withers and legs, lined-backed in color matching zebra markings.

3. *Pangaré.* (North Americans have no similar term). Any horse with a lighter, or faded, color area around the nose, eyes, belly, flanks, and genitals. The tops and sides of the body are a darker hue than the rest.

4. *Overo.* A horse with odd, irregular patterns of color. The face is generally oddly marked. The mane and tail may be any color or colors. Roan hairs on the body, mane, or tail are common.

5. *Pintado.* Any spotted horse, normally not an *overo* or a *tobiano.*

6. *Rodado.* A dappled horse.

7. *Rosillo.* A red roan. The term is also used as adjective when red hairs are found in other coat colors.

8. *Ruano.* A horse with any two or more colors mixed in the coat. Like *rosillo*, it is a modifying adjective.

9. *Sabino.* A horse with small clusters or spots of red hairs.

10. *Tobiano.* A horse with large areas of clearly defined color, white with some other base color. The face and legs are almost always normally marked. The mane and tail are one color, or, if two colors are present, they are independent with a clearly marked dividing line. Roan hairs are the exception.

Using the above terms as a starter, anyone knowing some Spanish should be able to talk about horses with a Latin-American horseman and understand or explain horse colors.

Bibliographical Sketch

The Arrival and Spread of the Horse in America

Because of the very nature of the subject, the materials covering the arrival and spread of the horse in America are scattered throughout all our records of the white man's conquest and expansion in the New World. Any commonplace essential such as the horse usually evoked only a passing reference from the historian, even though the success of the European in his various enterprises against the geographical and aboriginal barriers in the western Hemisphere can be largely attributed to his horse.

Manuscripts used in the preparation of the book, when unbound and not available to the general public, are not regularly indicated in the bibliography. Only books which are available to persons having access to the proper libraries are cited.

Background

The prologue containing an outline of the prehistoric horse in America is of necessity extremely brief. For a more authoritative and complete picture, one should read Richard Swann Lull's *The Evolution of the Horse Family* (New Haven, 1931) or W. D. Matthew and S. Harmsted Chubb's *Evolution of the Horse* (American Museum of Natural History, *Guide Leaflet Series No. 36*, New York, 1932).

The horse and the horsemanship of the Spaniard at the time of his arrival in the New World were principally of Moorish origin, and to understand the method and manner of training, riding, and fighting employed by the *conquistadores*, it is necessary to examine some European sources. Several Spanish, Italian, and Portuguese writers have devoted themselves to the noble equestrian art, and thus we have such priceless works as Don Bernardo de Vargas Machuca's *Libro de exercicios de la gineta* (Madrid, 1500), and Gonzalo Argote de Molina's *El Libro de la Montería* (Sevilla, 1582). Other works, such as Josep Delgado Hillo's *Tauromaquia ó arte de Torear á Caballo y á Pie* (Madrid, 1804) and Antonio Galvani de Andrada's *Arte cavallaria de gineta e estradiota* (Lisbon, 1678), are unexcelled for the Continental manner of horsemanship. There are four more books that could well be listed, namely, Suárez de Peralta's *Tractado de la Caballería de la Gineta y Brida* (Sevilla, 1580?); Cesare Feaschi's *Trattato Dell Imbrigliare, Atteggiare & Terrare Cavilli* (Venetia, 1603); Pedro Fernández de Andrada's *De La Natvraleza de Cavallo* (Sevilla, 1580); and Manuel Álvarez Assorio y Vega's *Manejo Real en que se propone lo que deben saber los Cavalleros . . .* (Madrid, 1764).

Works containing references to actual transport of horses to America are few in number. It is only by scrutinizing carefully documentary sources such as Martín Fernández de Navarrete's *Colección de los viajes y descubrimientos . . .* (Madrid, 1825–37, 5 vols.), and Pacheco y Cárdenas' *Colección de documentos inéditos relativos al descubrimiento . . .* (Madrid, 1864–84, 42 vols.), and piecing together the scraps of pertinent information from these extensive collections that any representative reconstruction can be made. In Ruidíaz y Caravia's *La Florida, su conquista y colonización por Pedro Menéndez de Áviles* (Madrid, 1893) is found one of the rare descriptions of how the horses were carried in the ships.

Conquest

Most of the contemporary historians of the conquest, as a rule, made only passing references to the horses, although their works must be read to obtain the story. Francisco López de Gómara's *La historia general de las Indias* (Antwerp, 1554), Antonio de Herrera y Tordesillas' *Historia general . . .* (Madrid, 1601–15, 8 vols. in 4), the same author's *Descripción de las Indias occidentales . . .* (Madrid, 1726), Gonzalo Fernández de Oviedo y Valdés' *Crónica de las Indias* (Salamanca, 1570), the same author's *Historia general de las Indias* (Madrid, 1851–55, 4 vols.), and a

Gentleman of Elva's *Relaçao Verdadeira dos Trabalhos do Governador Don Hernando de Soto* (Evora, 1557), are all basic. José de Acosta, in his *Historia Natural y Moral de las Indias . . .* (Sevilla, 1590, 2 vols.), speaks several times of the rapid growth of livestock. Speaking of the islands, on page 65, Vol. I, he says. "*Y ahora tienen innumerables manados de caballos, de bueyes, y vacas, de perros, de puercos . . . .*" Published collections of documents are tedious but essential, for example *Colección de los viajes y descubrimientos . . .*, edited by Martín Fernández de Navarrette (Madrid, 1825–37, 5 vols.), and the *Recopilación de documentos inéditos relativos al descubrimiento, conquista . . .* of Pacheco y Cárdenas (Madrid, 1864–84, 42 vols.)

Certain chronicles, however, prove the exception and do include many details concerning the horses. In works such as the Inca Garcilaso de la Vega's *La Florida del Inca* (Madrid, 1723), his *Primera parte de los comentarios reales* (Madrid, 1723), Agustín de Zárate's "*Historia del descubrimiento y conquista de la provincia del Perú*" (in *Biblioteca de Autores Españoles*, Madrid, 1826), and in Félix de Azara's *Apuntamientos para la historia natural de los cuadrúpedos del Paraguay y Rio de la Plata* (Madrid, 1802–1805, 3 vols.), numerous references to horses are found.

By far the outstanding work on the conquest, especially in the eyes of a horseman, is Bernal Díaz del Castillo's *Historia verdadera de la conquista de la Nueva España* (best edition edited by Genaro García, Mexico, 1904, 2 vols.); Díaz del Castillo, an unusual horseman in an age of *caballeros*, never missed an opportunity to tell of horses and, with the possible exception of the letters of Cortés (best English version probably F. A. McNutt's *Letters of Cortés*, New York, 1908, 2 vols.), his writings show his love of horses and horsemanship more than those of any other writer of the period. Also of interest is William Robertson's *The History of America* (London, 1777) and William H. Prescott's *History of the Conquest of Mexico* (New York, 1936). Modern writers, with two notable exceptions, have not been very much interested in the horses of the conquest. In Robert Bontine Cunninghame Graham's *Horses of the Conquest* (London, 1930) is found the best single account, though Graham occasionally allows his narrative powers to reconstruct the story beyond the known factors. His books on De Soto, Pedro de Valdivia, Bernal Díaz del Castillo, and the horses of the conquest are "must" books for horse-lovers. Outside of Díaz del Castillo and Graham, the best writer on the horses of the Conquest of Mexico is Frederico Gómez de Orozco who wrote the excellent treatise "*Los Cavallos de los Conquistadores*" (Sociedad Científica Antonio Alzate, *Memorias* Vol. XXXIX, Mexico, D.F., 1920–21).

Further sources are available. *Colección de libros y documentos referentes a la historia de America* (Madrid, 1904–29, 21 vols.) is good for the settlement of Argentina and Paraguay, especially Vols. V and VI. Attention is also called to Rui Díaz de Guzmán's *Historia de lo Rio de La Plata* (written in 1612, published in Asunción in 1845).

Expansion

The expansion of the horse into North America is rather well documented, although the narrative has never been brought together in one place.

Collections on Mexico such as Joaquin García Icazbalceta's *Colección de documentos para la historia de México* (Mexico, 1858–66, 2 vols.) and G. García and C. Pereyra's *Documentos inéditos ó muy raros para la historia de México* (Mexico, 1905, 35 vols.) are good. Many of the church histories, such as those written by J. Arlegui, F. X. Alegre, A. Pérez de Ribas, and A. Dávila Padilla are helpful, as is Matías de la Mota Padilla's *Historia de la conquista de la provincia de la Nueva Galicia* (Mexico, 1870). In later days many Mexican periodicals and papers are available. *El Museo Mexicano* and the *Gazeta de México* may be mentioned as examples.

For the advance into the trans-Mississippi West works such as Charles Wilson Hackett's *Historical Documents Relating to New Mexico, Nueva Viscaya and Approaches Thereto* (Washington, 1923–27, 3 vols.), Alonzo de Benavides' *Memorial . . .* (Madrid, 1630, *reimpreso*, Mexico, 1899), and Pierre Margry's *Memoires et documents pour servir a l'histoire . . .* (Paris, 1879–88, 6 vols.) are invaluable. Hubert Howe Bancroft's *Arizona and New Mexico* (San Francisco, 1885, 7 vols.) furnishes a good secondary source, and Herbert Eugene Bolton's *Rim of Christendom* (New York, 1936) and *Outpost of Empire* (New York, 1921) are excellent.

The acquisition of the horse by the Indians of North America has been outlined by Clark Wissler in his article entitled "The Influence of the Horse in the Development of Plains Culture" (*American Anthropologist,* Vol. XVI, No. 1 [1914]) and improved on by Francis Haines in the same review some years later (Vol. XL, No. 1 [1938]). Another important work is *The Indian and the Horse,* by Frank Gilbert Roe (Norman, University of Oklahoma Press, 1951).

For the Atlantic coast region, Genaro García's *Dos antiguas relaciones de la Florida* (Mexico, 1902), Herbert Eugene Bolton's *Spain's Title to Georgia* (Berkeley, 1925), Justin Winsor's *Narrative and Critical History of America* (New York, 1889, 8 vols.), and Reuben Gold Thwaite's *Jesuit Relations and Allied Documents* (Cleveland, 1896–1901, 73 vols.) are outstanding. Uncovering material on the horse before the eighteenth century necessitates, in almost every case, access to original narratives and correspondence from which the story must be pieced together.

There are a multitude of sources for the study of the horse on the Great Plains and the Pacific Slope. Journals of early travelers represent possibly the best source. Fortunately for those interested, the best of these have been edited by Reuben Gold Thwaites in his many-volumed work entitled *Early Western Travels 1748–1846* (Cleveland, 1904–1907, 32 vols.). Particular attention is drawn to James Ohio Pattie's *Personal Narrative of a Voyage to the Pacific and in Mexico* (Vol. XVIII), to Josiah Gregg's *Commerce of the Prairies* (Vol. XIX), to Thomas Jefferson Farnham's *Travels in the Great Western Prairies* (Vol. XXIX), and to Colonel Henry Inman's *The Old Santa Fé Trail. Niles Weekly Register* (1811–49, Baltimore, Ohio) is also a mine of information, as are *The Works of Hubert Howe Bancroft* (San Francisco, 1882–90). For those interested in early California horses and horsemanship, particular attention is drawn to Bancroft's volume entitled *California Pastoral.* For a general survey of the Great Plains and its occupation by the Anglo-Americans, Walter Prescott Webb's *The Great Plains* (Boston, 1931) is unsurpassed. For an overview, J. Frank Dobie's *The Mustangs* (Boston, 1952) is an excellent selection.

For the movement of the horse into South America there is an equally large group of material and some specialized studies. Best of this later group are works such as "The Spanish Horse of the Pampas" (*American Anthropologist*, Vol. XLI, No. 1 [1939]) and *"El Caballo Argentino"* (Boletín del Ministerio de Agricultura, Buenos Aires, 1900) and "*El Caballo Criollo*," by Desiderio Davel (*Boletín* del Ministerio de Agricultura, Buenos Aires, 1921); and "*Origin de la Boleadora y del Caballo en la República Argentina*," by Anibal Cardosa (*Anales* de Museo Nacional de Historia Natural, *Tomo* XXVIII, Buenos Aires, 1916). Introduction of livestock in general, and particularly horses, has also been treated by others. Examples are: Enrique Lynch Arribalzaga "*Origin y Caracteres del Caballo Criollo*," (*Anales*, Ministerio de Agricultura, Buenos Aires, 1900); Erland Nordenskiöld *Deductions Suggested by the Geographical Distribution of some Post Colombian Words used by the Indians of South America.* (*Comparative Ethnographical Studies*, Goteberg, 1919–31, 9 vols.). Dr. Prudencio de la Mendoza's *La Ganadería Colonial del Siglo XVI* (Buenos Aires, 1928) is also good.

One of the best histories of the South American horse is found in a book entitled *El Caballo Chileno* written by Uldaricio Prado (Santiago, 1914). The story of the introduction of horses into South America is covered briefly in *Trabajos Geográficos de la Casa de Contratación* (Sevilla, 1900), by Manuel de la Puente y Olea, and Ricardo Cappa's *Estudios Críticos*, Vol. IV (Madrid, 1889–, 15 vols.) is especially good, as is also Bernarbé Cobo's *Historia del Nuevo Mundo* (Sevilla, 1890–95, 4 vols.), whose forty years' residence and acquaintance with early settlers gave him excellent opportunities to acquire pertinent knowledge. The utilization of the horse on South American ranches is clearly reflected, for Argentina at least, in *Instrucciones á las Mayordomas de Estancias*, by Juan Manuel de Rosas (Buenos Aires, 1942), who gained his greatest fame as a political leader and dictator but who also showed that he knew the work of a Gaucho. A guide to the expansion of the horse on the eastern side of the La Plata is found in *La Difusión del Bovino en Nuestro Uruguay*, by B. Caviglia (Montevideo, 1935).

Basic source material is found in such contemporary accounts as Thomas Gage's *A New Survey of the West India* (London, 1648, under a much longer title) and in Alonso de Ovalle's *Histórica relación del reyno de Chile . . .* (Roma, 1646); and in later works such as Edward Gaylord Bourne's *Spain in America 1450–1580* (New York, 1904), and in Enrique de Gandia's *Historia de la Conquista del Rio de La Plata y del Paraguay* (Buenos Aires, 1932). There is a scarcity of good books covering Brazilian history, but Pedro de Magalhães' *This Histories of Brazil* (originally published in 1576) is one of the best. Also good are *Historia da Civilização Brasileira* (São Paulo, 1940), by Pedro Calmon; *History of Brazil*, by Robert Southey (London, 1822, 3 vols., second edition); and *A History of Brazil*, by João Pandia Calogeras, translated by Percy Alvin Martin (Chapel Hill, 1930).

The Western Horse Today

Beginning about the end of the nineteenth century, the available materials on the Western horse are greatly increased. Robert Cunninghame Graham, mentioned above, has written numerous essays and books which include material on the Western horse of both North and South America. For material on the Venezuelan cowboy, read his *José Antonio Páez* (London, 1929). João do Norte (pseudonym of Gustavo Barroso) occupies a similar position, although he confined his writings to his native Brazil. His works are a mine of customs, legends, horse and folk lore. His best work is probably *Terra de Sol* (Rio de Janeiro, 1912). More scientific in his approach but just as interesting as Justo P. Sáenz (*hijo*) who wrote for Argentina *Equitación Gaucho* (Buenos Aires, 1942). The best single volume bringing together the Western horse lore of North America is J. Frank Dobie's (ed.) *Mustangs and Cowhorses* (Austin, 1940).

A number of contemporary magazines and reviews have been concentrating on articles dealing with the different Western types and breeds and their history in general. The best of them include *The Western Horseman* (Colorado Springs, Colo.), *Anales de la Asociación Criadores de Criollo* (Buenos Aires, Argentina), *Anais de Associąção dos Criadores de Cavallos Crioulos* (Pelotas, Brazil), and *The Cattleman* (Fort Worth, Texas).

For excellent material on the history of the development of the Quarter Horse, see Patrick Nesbitt Edgar's *The Sportsman's Herald and Studbook* (New York, 1833), Sanders D. Bruce's *American Studbook* (New York, 1873), and John H. Wallace's *The Horse of America* (New York, 1897). See also *The Quarter Horse* (Fort Worth, 1951), and *Quarter Horses: A Story of Two Centuries*, by Robert Moorman Denhardt (Norman, University of Oklahoma Press, 1967). Another good book is *America's Quarter Horses*, by Paul Laune (New York, 1973).

For data concerning the Palomino horse publications issued by the Palomino Horse Breeders of America, Inc. (Melba Lee Spivey, executive secretary, Mineral Wells, Texas) are available; particular attention is drawn to their publication *Palomino Horses* (monthly). The Appaloosa (Appaloosa Horse Club, George B. Hatley, executive secretary, Moscow, Idaho) is not wanting for literature. There are available more than a dozen studbooks, as well as general-interest books, pamphlets, and films. The best single book is *The Appaloosa: The Spotted Horse in Art and History*, by Francis Haines (Austin, 1972). The Albino Horse Association (Ruth Thompson White, secretary, Crabtree, Oregon) has several descriptive phamplets, such as *The American White Horse*, which are worthwhile. Both the Pinto and the Paint horse associations have literature available. Write the Pinto Horse Association of America (San Diego, California), for information about their horses, and write the American Paint Horse Association (Sam Ed Spence, executive secretary, Fort Worth, Texas) for an excellent booklet entitled *Paints*. The Colorado Ranger headquarters is in Litchfield, Ohio. The Spanish Mustang Registry (Mrs. Leana Rideout, executive secretary, Marshall, Texas) is also willing to provide information. Other breed organizations are located as follows: the Spanish Barb Mustang Breeders Association (Peggy Cash, secretary, Colorado Springs, Colorado); the International Buckskin Horse Association (Richard Kurzeja, executive secretary, St. John, Indiana); the American Buckskin Registry Association (Marilyn Johnston, executive secretary, Anderson, California); the American Indian Horse Registry (Larry Klutt, secretary, Apache Junction, Arizona); the American Paso Fino Horse Association (Mary Lohrentz, registrar, Pittsburg, Pennsylvania).

For data relative to the *Criollo* and *Crioulo* horses of Argentina, Uruguay, and Brazil, see the publications of the various associations (listed below) and their studbooks. Also good is *Vocabulario y Refranero Criollo* by Tito Saubidet, which is a most beautiful book, illustrated by the author. It is a *criollo* and *estancia* dictionary.

The studbooks, when printed, are excellent sources of historical and factual information relative to South American breeds; for example, *Stud Book de la Raza Chilena* (Sociedad Nacional de Agricultura, Santiago, Chile) and *Stud Book Argentino para la Raza Criollo Argentina* (Sociedad Rural Argentina, Buenos Aires) and *Official Stud Book and Registry of the American Quarter Horse Association* (American Quarter Horse Association, Amarillo, Texas). The present secretary of the American Quarter Horse Association is Don Jones, P.O. Box 200, Amarillo, Texas.

There are not many books on the modern horse in Brazil, but two which could be mentioned are *O Cavallo Crioulo*, by D. M. Riet (Porto Alegre, 1918); and *Contribuição para o Estudo da Criação do Cavallo,* by P. de Lima Correa (São Paulo, 1935).

The *charro* of modern Mexico is covered in an interesting if unscientific fashion in *El Charro Mexicano*, by Carlos Ricon Gallardo (Mexico, 1939), and rather well in *Historia de la Charrería* by José Álvarez del Villar (Mexico, 1941).

Index